my **revision**

GET BETTER RESULTS FOR AQA

AQA GCSE
BIOLOGY

For A* to C

Mike Boyle

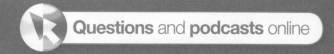

PHILIP ALLAN
UPDATES

Philip Allan Updates, an imprint of Hodder Education, an Hachette UK company, Market Place, Deddington, Oxfordshire OX15 0SE

Orders

Bookpoint Ltd, 130 Milton Park, Abingdon, Oxfordshire OX14 4SB

tel: 01235 827827

fax: 01235 400401

e-mail: education@bookpoint.co.uk

Lines are open 9.00 a.m.–5.00 p.m., Monday to Saturday, with a 24-hour message answering service. You can also order through Philip Allan Updates website: www.philipallan.co.uk

© Mike Boyle 2011

ISBN 978-1-4441-2083-7

Impression number 5 4 3 2 1

Year 2016 2015 2014 2013 2012 2011

Cover photo reproduced by permission of XYZproject/Fotolia
Other photos are reproduced by permission of the following: **p25** DoctorKan/Fotolia; **p37** alessandro merlini/Fotolia

Printed in Spain

Hachette UK's policy is to use papers that are natural, renewable and recyclable products and made from wood grown in sustainable forests. The logging and manufacturing processes are expected to conform to the environmental regulations of the country of origin.

Get the most from this book

This book will help you revise units Biology 1–3 of the new AQA specification. You can use the contents list on pages 2 and 3 to plan your revision, topic by topic. Tick each box when you have:

1 revised and understood a topic

2 tested yourself

3 checked your answers and practised exam questions online

You can also keep track of your revision by ticking off each topic heading through the book. You may find it helpful to add your own notes as you work through each topic.

Tick to track your progress

examiner tips

Throughout the book there are exam tips that explain how you can boost your final grade.

Higher tier

Some parts of the AQA specification are tested only on higher-tier exam papers. These sections are highlighted using a solid yellow strip down the side of the page.

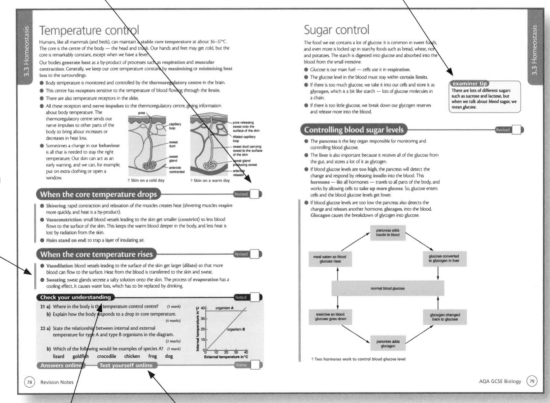

Check your understanding

Use these questions at the end of each section to make sure that you have understood every topic.

Go online

Go online to check your answers at **www.therevisionbutton.co.uk/myrevisionnotes**.

Here you can also find extra exam questions for topics as well as podcasts to support you when getting ready for the big day.

Contents and revision planner

B3 Biology 3

Diet and exercise

A balanced diet

A balanced diet contains the right amount of energy and a balance of different nutrients:

- **Carbohydrates** (sugars and starch) for energy.
- **Fats** (lipids) to make certain components of cells, such as **membranes**. Fat storage tissue is important for **insulation**, body shape and **protection** of internal organs.
- **Proteins** for growth, cell repair and cell replacement.
- Small amounts of **minerals** and **vitamins** for various chemical reactions in the body.
- **Fibre** to allow the gut to push food along efficiently. Most fibre is plant material that we cannot digest.

1 People who are starving are **undernourished**. As well as being underweight, the lack of food can cause a **reduced resistance to infection**. Children fail to grow and develop normally, and women may have **irregular periods**.

2 People who do not eat a balanced diet are **malnourished**, and their health will suffer in some way. In the developed world, people are more likely to be overweight than underweight.

3 A lack of a particular vitamin or mineral will cause a **deficiency disease**; for example, a lack of vitamin D can cause rickets, a condition in which the bones do not develop properly.

↑ **If you want to stay healthy, five words of advice: eat plants, exercise, don't smoke**

Balancing energy input and output

A key idea here is that of **metabolic rate**:

- **Metabolism** is the general term for the chemical reactions inside your body.
- **Metabolic rate** refers to the speed or rate of these reactions — it is basically the same as the rate of **respiration**, because respiration provides the energy that all the other reactions need.

People with a **high metabolic rate** will tend to be thin, and can eat a lot of food without putting on weight. People with a **low metabolic rate** tend to be overweight, and a higher proportion of their food will be stored as fat.

The amount of energy you need from food depends on:

- your **age** — metabolic rate tends to slow down as we get older
- your **sex** — boys tend to have a higher proportion of muscle to fat than girls do, and muscle has a higher metabolic rate
- your **genes** — some people **inherit** a tendency to have a high or low metabolic rate, or a certain muscle-to-fat ratio

When we exercise, our metabolic rate **stays high** for quite a while after exercise, so exercise is a great way to control weight.

How much energy do we need?

- All of us need a certain amount of energy, which we get from our food.
- Proteins, carbohydrates and fats **all contain energy**.
- This energy is used for many different reactions in the body, including those involved in the movement of muscles and keeping warm.
- **Exercise** (moving muscles) increases the amount of energy expended by the body.

The only way to lose weight is to change the input/output balance, which means to reduce the energy content of your diet and be more active. Claims such as 'eat yourself slim' are unlikely to have any scientific basis.

> **examiner tip**
>
> Respiration does not make energy from food. You cannot make energy. Respiration *releases* energy from organic molecules such as sugars and lipids.

Check your understanding

1 Match the terms below to parts **a)**–**c)** in the paragraph.   *(3 marks)*

   glucose   metabolic rate   energy requirements

   Of all sporting events, the Tour de France is one of the most demanding. Cyclists are on their bikes for long periods and for many days, so their _____**a)**_____ will be very high. It has been estimated that their daily energy needs are 30 000 kJ, while the average male of the same age needs about 12 500 kJ. The cyclists take regular drinks containing _____**b)**_____, our body's immediate source of energy. Being so active, these athletes will have a high _____**c)**_____ .

2 The labels on a chocolate bar state that it contains 1000 kJ. The table shows the energy expended by an average 25 year old female office worker. For how long would the individual need to be doing the following activities in order to work off the chocolate bar?

   a) swimming

   b) sitting, watching television   *(2 marks)*

Activity	Energy used in kJ/min
Standing, cooking	9
Sitting, watching television	5
Walking briskly	15
Climbing stairs	28
Swimming	35
Dancing	19
Jogging	27

Fighting infectious disease

There are two types of disease:

1 **Infectious** — diseases that you can catch. These are caused by **pathogens**, which are **microorganisms** that cause **disease**. The main pathogens are **bacteria** and **viruses**.

2 **Non-infectious** — diseases you cannot catch. These are caused by other factors, such as **lifestyle** (diet, smoking, etc.), your **genes**, or simply **getting older**.

Bacteria
Revised

- Bacteria are a major cause of disease.
- They are tiny, single-celled organisms. Up to 1000 or more bacteria can fit inside one of your cells. They can be viewed with a light microscope.
- Only a few types of bacteria cause disease. Most are harmless and many are useful.
- Bacteria cause disease when they multiply inside our bodies.

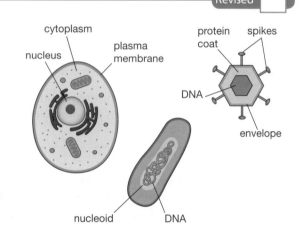

→ **Simple diagrams of a human cell (left), a bacterial cell (centre) and a virus (right); not to scale – the bacteria and viruses are much smaller compared with the human cell**

Viruses
Revised

- Viruses are so small that even with the best light microscope you cannot see them — you need an electron microscope.
- Viruses cannot reproduce on their own. They reproduce by getting inside your cells and using their own DNA to instruct the cell to make more viruses. The new viruses burst out of the cell and go on to infect more cells.

Diseases caused by bacteria	Diseases caused by viruses
Food poisoning	HIV/AIDS
Cholera	Influenza
Typhoid	Colds
Leprosy	Chicken pox

The **symptoms** of disease are caused in two ways. Bacteria and viruses produce **toxins** that interfere with the normal functioning of the body. Viruses also damage our cells when they burst out. Painkillers like paracetamol will only make you feel better, but will not get rid of the bacteria or viruses.

When microbes get inside the body
Revised

This is when the **immune system** is needed. The first job is to tell the difference between our own cells and any invaders. We have a number of different types of **white blood cells** whose job is to **recognise** these 'foreign' pathogens, by the chemicals, or **antigens**, on their outer surface.

1 Once a pathogen has been identified, some white blood cells will make **antibodies** that attach to the pathogen so that it can be destroyed.

2 The body can make thousands of different antibodies, tailor-made for particular pathogens. So if the chicken pox virus gets into our blood, we can respond by making anti-chicken pox antibodies.

3 Some white blood cells produce **antitoxins**, which neutralise the toxins (poisons) released by pathogens.

4 Some white blood cells **ingest** (engulf) pathogens or take them into the cell, where they are destroyed.

Vaccines prepare the immune system
Revised

When the body meets a disease for the first time, it will have no antibodies for that disease, and so will be vulnerable. The problem is that you cannot make antibodies before you have been exposed to the disease. **Vaccination** or **immunisation** is used to protect against particular deadly diseases **before** infection.

A **vaccine** contains something that will **stimulate** the immune system to make the right antibodies without causing the disease. Vaccines can be made with **dead pathogens**, live but **weakened** (harmless) pathogens, or **purified antigens**. If the pathogen gets into the body of a vaccinated person, that person will be **immune** to the pathogen. The more people vaccinated against a particular disease, the less likely the disease is to spread.

Vaccines have saved millions of lives. However, in rare cases vaccines may have serious **side effects**. In the late nineties a study claimed a link between the measles, mumps and rubella (**MMR**) vaccine and **autism** in children. Some parents opted not to give the vaccine to their children. The controversy has since died down because the claim was found to be grossly misleading.

It is all about weighing up the **risks**. Ask yourself:

● what are the chances of getting the disease?
● what are the chances of being damaged by the vaccine?

With measles the chances of getting complications like pneumonia are about one in 15, and about one in 500 die from the disease. The chance of having a problem with the vaccine is about one in 1 million.

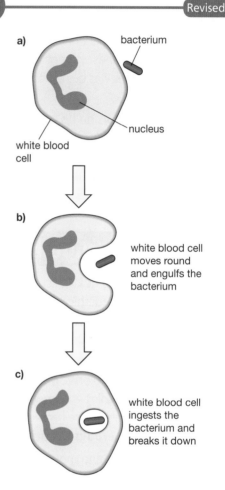

↑ A white blood cell engulfing and destroying a bacterium — enzymes are discharged in the vacuole of the white blood cell, and these break down the bacterium

Check your understanding
Tested

3 State the three ways in which white blood cells defend against pathogens. *(3 marks)*

4 Does every single person need to be vaccinated in order for that vaccine to be effective? Explain. *(3 marks)*

5 Evaluate the claim that 'all children should vaccinated against measles'. *(2 marks)*

examiner tip

'Evaluate' means 'weigh up' or 'look at the good points and bad points'.

Answers online — **Test yourself online**
Online

The fight against disease: then and now

Two hundred years ago, the treatment of infectious diseases left a lot to be desired. People had never heard of bacteria and viruses, and the need to be clean and avoid infection was not understood. Hospitals were dangerous places.

Semmelweis
Revised

In 1840, in Vienna Hospital, **Ignaz Semmelweis** noticed that a lot of pregnant women (about 12%) were getting infections and dying soon after childbirth. He noticed that doctors would examine the women straight after treating other patients, or even after examining dead bodies. The need to wash their hands was not appreciated. Semmelweis said that material from dead bodies was being transferred and was causing the disease.

Semmelweis told the doctors on his ward to **wash their hands** in **antiseptic**, and within a short time the death rate fell to 2%.

The antiseptic was killing bacteria, although the doctors and scientists did not know that. The link between bacteria and disease was not established for another 20 years.

Antibiotics and 'superbugs'
Revised

Antibiotics are medicines that kill bacteria inside the body. Common examples include penicillin, methicillin and erythromycin. They cannot kill viruses. Before antibiotics were discovered, bacterial infections could not be treated, so the mortality rate was very high.

> **examiner tip**
> Don't confuse anti*biotics* with anti*bodies*.

Owing to **over-use** of antibiotics, some strains of bacteria are now resistant to one or more antibiotics. **MRSA** stands for methicillin-resistant *Staphylococcus aureus*, meaning that methicillin is no longer effective against that particular strain of the bacterium *Staphylococcus aureus*. The media gave the name **superbugs** to **antibiotic-resistant bacteria**.

Disturbingly, there are more and more strains of bacteria getting resistant to more and more antibiotics. If no antibiotics were effective, we would be back to the days before penicillin — no treatment and a high death rate.

The good news is that we are developing new antibiotics all the time. Doctors also do not use antibiotics to treat minor infections such as sore throats. This slows down the rate of development of new strains of resistant bacteria.

Antibiotic resistance
Revised

There are three steps to antibiotic resistance:

1 Bacteria exist in vast numbers, and there is variation in the population.

2 Some bacteria are naturally resistant to penicillin. This could be due to a chance **mutation**, or it may be due to a gene that the bacteria have had for a long time.

3 The resistant **strain** of bacteria **survives** and **reproduces**, passing its resistance genes on to the next generation. This is **natural selection**. The population of resistant **bacteria** increases.

People can unwittingly speed up the evolution of resistant bacteria by not following the instructions given to them with their antibiotics.

● We only get symptoms when the bacterial population (in your throat, for example) reaches a certain size.

● When we begin to take antibiotics, the least resistant bacteria die first.

● When the population falls below the symptoms threshold, we often think we are cured and do not bother taking the whole course.

● What we have done is to let the most resistant bacteria 'off the hook' so they can reproduce and create a new generation of even more resistant bacteria.

Mutations and epidemics ──────────────── Revised

Mutations occur in both bacteria and viruses. In viruses, a new mutation could make the current vaccine ineffective. This happens every year with different strains of the flu virus — a **new vaccine** has to be developed to fight the new virus strain.

If a new vaccine is not developed quickly, the disease will spread rapidly because people are not immune to it. A widespread outbreak is called an **epidemic**. An epidemic that spreads across continents is called a **pandemic**.

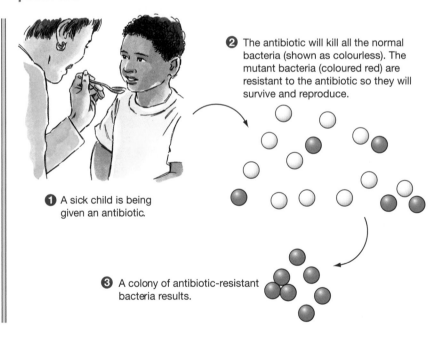

❶ A sick child is being given an antibiotic.

❷ The antibiotic will kill all the normal bacteria (shown as colourless). The mutant bacteria (coloured red) are resistant to the antibiotic so they will survive and reproduce.

❸ A colony of antibiotic-resistant bacteria results.

← Development of a strain of bacteria resistant to an antibiotic

examiner tip

Many students imply that the bacteria 'decide' to become resistant. Organisms cannot do this. They were 'born' lucky — with the right combination of genes.

Check your understanding ──────────────── Tested

6 Explain why it is important to avoid over-use of antibiotics. *(2 marks)*

7 Suggest what steps should be taken in hospitals to reduce the spread of MRSA. Use the information about Semmelweis to help you. *(2 marks)*

8 Why don't doctors prescribe antibiotics to treat colds and flu? *(1 mark)*

examiner tip

If the exam question asks you to 'use the information', you must base your answer on the information provided.

Answers online ──── **Test yourself online** ──────────────── Online

The nervous system

The nervous system

1 The nervous system allows you to **detect** what is going on both inside and outside your body, and to **respond** in the right way.

2 It does this by detecting **stimuli** and sending this information into your **central nervous system (CNS)**, which is your **brain and spinal cord**.

3 A **stimulus** is a change in the environment that we can detect, such as light, sound, movement, heat, etc.

4 **Receptors** are cells that can detect a particular stimulus. **Sense organs**, like the eye, ear and skin, contain a lot of these receptor cells.

5 Receptor cells send this information along specialised cells called **neurones**, or **nerve cells**, which are like tiny wires.

6 Most of this sensory information goes to our **brain**, which **processes** the information and **coordinates** a response, often based on memory. For instance, we may see somebody, recognise them, and call out their name.

7 The organ that makes the response is called the **effector**. This could be a muscle, which responds by contracting, or a gland, which responds by releasing a chemical, for example a hormone.

examiner tip

Many exam answers confuse sense organs with receptors — 'the skin is a receptor' would get no marks. The skin is a sense organ that contains many different receptor cells.

A single neurone or nerve cell

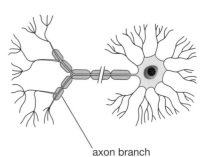

axon branch

⬆ **Neurones (or nerve cells) can be very long and transmit information from one part of the body to another**

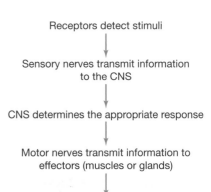

Receptors detect stimuli
↓
Sensory nerves transmit information to the CNS
↓
CNS determines the appropriate response
↓
Motor nerves transmit information to effectors (muscles or glands)
↓
Effectors bring about response

⬆ **Overview of the nervous system**

The senses

There are five senses. They are:

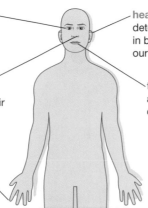

sight — receptors in our eyes are sensitive to light so we can detect colour and shape

smell — receptors in the nose detect thousands of different chemicals in the air that go up our nose

touch — receptors in our skin are sensitive to light and heavy touch (pressure), pain and temperature

hearing — receptors in our ears detect sound as well as changes in body position that help us keep our balance

taste — receptors in our tongue are sensitive to chemicals and can detect five basic tastes

Reflexes

A **reflex** is a quick response. There are many different reflexes, often designed to protect us from danger, or to keep our balance and body position. Examples of reflexes include:

● blinking when something comes near to your eye
● jerking your hand away from a hot or sharp object
● the knee jerk (when tapped just under the kneecap)

The key features of reflexes are:

● they are fast
● they are automatic — they do not involve conscious control by the brain, so you cannot stop them, even if you wanted to
● the **same stimulus** always leads to the **same response**
● they involve only three neurones: sensory, relay and motor

↓ **The components of a reflex action**

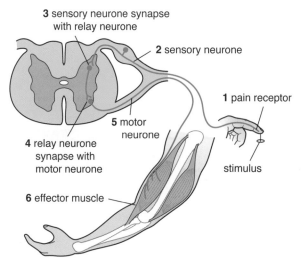

3 sensory neurone synapse with relay neurone
2 sensory neurone
1 pain receptor
5 motor neurone
4 relay neurone synapse with motor neurone
stimulus
6 effector muscle

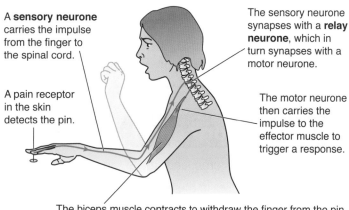

A **sensory neurone** carries the impulse from the finger to the spinal cord.

A pain receptor in the skin detects the pin.

The sensory neurone synapses with a **relay neurone**, which in turn synapses with a motor neurone.

The motor neurone then carries the impulse to the effector muscle to trigger a response.

The biceps muscle contracts to withdraw the finger from the pin. Note: the reflex does not pass through the brain but a separate neurone carries an impulse to the brain from the relay neurone so that you are aware of the reaction just after it has happened.

Synapses

Revised

1 A **synapse** is a junction between two neurones, such as between a relay neurone and a sensory neurone in a reflex action.

2 When a nerve impulse reaches a synapse, a **transmitter chemical** is released. This passes (diffuses) across the gap and sets up an impulse in the next neurone.

examiner tip

Always talk about nerve impulses, not messages or signals. A nerve impulse is a tiny electrical 'blip' and it is the brain's job to make sense of it and 'decode' the message.

→ A synapse

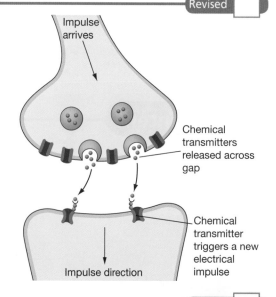

Impulse arrives

Chemical transmitters released across gap

Chemical transmitter triggers a new electrical impulse

Impulse direction

Check your understanding

Tested

9 a) List six different stimuli that you can detect. *(3 marks)*

  b) State two differences between sensory and motor nerves. *(2 marks)*

10 Match the terms below to parts **a)–e)** in the paragraph. *(5 marks)*

effectors   **motor nerves**   **sense organ**
sensory nerves   **receptor cells**

The eye is an example of a _____**a)**_____ that contains millions of _____**b)**_____, which are sensitive to light. The information they gather passes along _____**c)**_____ to the brain. The brain processes this information and coordinates a response. Impulses pass along _____**d)**_____ to the _____**e)**_____, which could be the muscles of the chest, throat and tongue that allow us to speak.

Answers online   **Test yourself online**

Online

Control in the human body

Homeostasis

Revised

Homeostasis means 'keeping constant internal conditions' — while your outside environment may change, the conditions inside your body stay remarkably constant. Homeostasis is usually achieved in three steps:

1 The change is detected by **receptor cells**, usually in the central nervous system. For example, a particular area of the brain is sensitive to a change in the temperature of the blood flowing through it.

2 The change is reversed. We shiver or sweat, for example.

3 The change is monitored and the mechanisms are switched off when no longer needed.

Homeostasis is achieved by a combination of nervous impulses and hormones.

What are hormones?

Revised

1 **Hormones** are compounds made and secreted by **glands** such as the ovaries, testes and pituitary gland.

2 They always travel in the blood.

3 They have an effect on **target cells**. These cells may be in one particular organ, or scattered throughout the body.

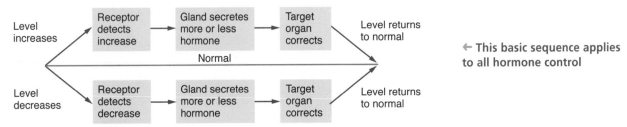

← This basic sequence applies to all hormone control

Water content

Revised

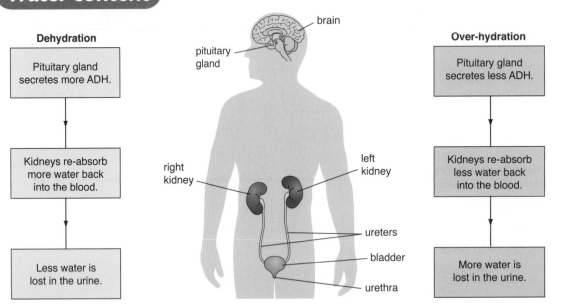

↑ The control of water balance involves one hormone called ADH (anti-diuretic hormone). Diuresis is the production of urine, so think of it as the 'stop you peeing' hormone

We do not want to become dehydrated or over-hydrated. We can lose water in several ways:

● in sweat   ● in our breath

● in urine   ● in faeces (we lose a lot more water when we have diarrhoea)

This water loss must be replaced from our food and drink.

Water balance is largely carried out by the **kidneys**. If we drink a lot and do not sweat, the kidneys will get rid of the excess water by producing lots of dilute urine. If we sweat a lot and do not drink, the kidneys will conserve water by producing a small amount of concentrated urine. If your urine is dark, you need to drink some water.

Ion content
Revised

Important ions include:

- sodium and chloride
  (sodium chloride = common salt)
- potassium
- calcium

These ions are sometimes called **salts**, or **electrolytes**. We get these ions in our diet, and the kidneys remove the excess. We are rarely short of electrolytes, but it can happen during severe **diarrhoea**, when the body does not have a chance to absorb the ions from the gut, or when we have lost a lot of **sweat**.

Temperature
Revised

- The control of temperature is vital because the chemical reactions of the body are controlled by enzymes, which work together most efficiently at about 37°C.

- Heat loss can be increased by **sweating** and by allowing blood to flow close to the surface of the skin. Heat loss can be lowered by reducing blood flow to the skin.

- Also, when we are too cold we **shiver** because the rapid movement of muscles produces heat. If we are cold for long periods, the body can also increase its **metabolic rate**.

→ Isotonic drinks are designed to replace the water and ions lost by sweating, and have added glucose for energy. The work isotonic means they have the same concentration as the fluids that were lost, i.e. the sweat

Blood sugar
Revised

- **Blood sugar** (glucose) levels need to be kept within certain limits in order to provide cells with a constant supply of **energy**. People who cannot control their blood glucose levels are **diabetic**.

- Much of our glucose comes from the digestion of carbohydrates such as starch. We can store glucose as **glycogen**, mainly in the **liver**.

- The control of blood glucose involves storing glucose when we have too much, and releasing it when there is not enough.

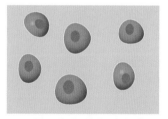

↑ All the cells in your body are surrounded by tissue fluid, from which they get all their nutrients and oxygen. They also excrete their waste into it. Overall, the aim of homeostasis is to keep all our cells bathed in just the right conditions

Check your understanding
Tested

11 If you have been exercising, you will have lost water and started to become dehydrated.
   a) State two ways in which exercise may increase water output. *(2 marks)*
   b) Give a reason why you should drink fluids containing both water and ions. *(1 mark)*
   c) Give an example of one other circumstance in which the body may need to replace both water and ions. *(1 mark)*

12 Suggest when a person's blood glucose levels will be:
   a) highest *(1 mark)*
   b) lowest *(1 mark)*

Answers online   Test yourself online   Online

The menstrual cycle and fertility

The menstrual cycle is a woman's monthly reproductive cycle. Every month an egg, or **ovum**, is developed in one of the **ovaries** and then released, in a process called **ovulation**.

There are three hormones involved: **FSH**, **oestrogen** and **LH**. There is a fourth hormone, **progesterone**, but its role in the menstrual cycle is not on the specification. However, you are expected to know that it is an important part of the contraceptive pill.

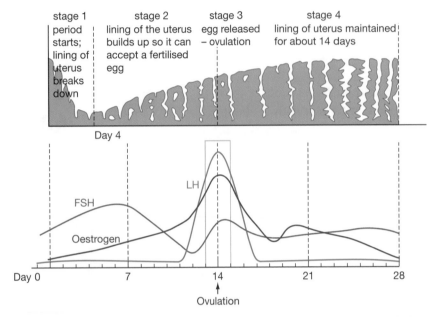

stage 1
period starts; lining of uterus breaks down

stage 2
lining of the uterus builds up so it can accept a fertilised egg

stage 3
egg released – ovulation

stage 4
lining of uterus maintained for about 14 days

Day 4

← The changes in the uterus lining and levels of oestrogen, LH and FSH during an average 28-day menstrual cycle

LH

FSH

Oestrogen

Day 0   7   14   21   28

↑
Ovulation

Hormone	Full name	Where it is made	What it does	When it acts
FSH	Follicle-stimulating hormone	Pituitary gland	**1** Stimulates the development of an egg inside a follicle **2** Stimulates the ovary to release oestrogen	Day 1 (when the period starts) to about day 13
Oestrogen	Oestrogen	Ovary	**1** Causes uterus lining to thicken **2** Inhibits production of more FSH **3** Stimulates the pituitary gland to produce LH	Day 5 (after period) to day 28 Day 7–8 Day 11–12
LH	Luteinising hormone	Pituitary gland	Causes ovulation — the release of an egg from an ovary	About day 14

As hormones control the development of eggs and ovulation, we can use hormones as **contraceptives**, and we can also use them in **fertility treatment** to help couples to conceive a baby.

1 The contraceptive pill ('**the pill**') contains oestrogen and **progesterone**. High levels of these hormones inhibit the production of FSH. No FSH means no egg develops, no ovulation and therefore no baby.

2 A common cause of **infertility** is **low levels of FSH**, so that eggs do not mature and cannot be released. Treatment involves taking FSH to **stimulate egg production**.

The pill has advantages and disadvantages:

- It is taken **orally**, which is easier and more pleasant than injections or implants.
- It is very **effective** — 99% of women on the pill do not get pregnant — but some do.
- It can make periods lighter and less painful.
- It gives no protection against sexually transmitted diseases, such as HIV and chlamydia.
- It can have **side effects**, such as headaches, nausea, irregular periods and water retention. It is also thought to increase the risk of **deep vein thrombosis** and **heart disease**.
- To minimise these risks, modern-day pills have lower doses of oestrogen, and there is also a **progesterone-only pill** that has even fewer side effects and a lower risk of heart disease.

IVF stands for **in vitro fertilisation**, where 'in vitro' means 'in glass'. The woman is given FSH so that her ovaries produce more eggs than normal, and LH to stimulate the release of these eggs. The eggs are collected and then fertilised by sperm in vitro. The fertilised eggs develop into embryos. A few healthy embryos are implanted into the woman's uterus.

Risks and problems of fertility treatment:

- It may not work, and trying again and again can be time-consuming, stressful and expensive.
- It can result in **multiple births** (triplets, quads etc.).

> **examiner tip**
>
> Make sure that you can list the ethical issues, as well as the benefits and risks of IVF. Many people object to IVF on the grounds that an embryo is a human being, with rights, from the moment of conception. IVF creates many embryos that will never be used, and this could be seen as a waste of human life.

Check your understanding

Tested

13 Match the terms below to parts **a)**–**d)** in the paragraph.   *(4 marks)*

ovaries   oestrogen   luteinising hormone   pituitary

Puberty in girls begins when the _____**a)**_____ gland begins to secrete the hormone FSH. The target organs of this hormone are the _____**b)**_____ . In turn, this gland secretes the hormone _____**c)**_____ . Rising levels of this hormone stimulate the pituitary gland to release the hormone _____**d)**_____ , which causes ovulation.

14 a) At what point in the menstrual cycle is a woman at her most fertile?   *(1 mark)*

b) A simple blood test can be used to diagnose reproductive hormone deficiencies in women. Explain why a blood test for low hormone levels must be done at a certain time of the month.   *(2 marks)*

Answers online   Test yourself online   Online

Control in plants

Plants have a few basic requirements:

- **light** for photosynthesis
- **water**
- **carbon dioxide** for photosynthesis
- Some **minerals** such as nitrate and phosphate
- a little **warmth**

Plants have no muscles and usually cannot move as quickly as animals, but they do need to detect and respond to several important stimuli in their environment. Usually, plant movement is slow because it is brought about by **cell division** and **elongation**.

Most plants can respond to **light**, **moisture** and **gravity**. A germinating seed in the ground needs to be able to detect which way is up, so that the roots can grow down and the shoots can grow up. Most seeds use gravity to achieve the correct growth shortly after germination.

Plant responses are called **tropisms**. A response to light is called a phototropism. Movement towards light is a **positive phototropism**; movement away is a **negative phototropism**.

Similarly, there are positive and negative **hydrotropisms** (water) and **gravitropisms** (gravity). Gravitropisms are sometimes called geotropisms.

← Plants grow towards a light source — this is positive phototropism

How do plants respond to light?

Revised

Phototropism is controlled by a hormone called **auxin**, which is produced in the tip of the plant. Auxin causes cells to divide and to get longer.

- If the plant is lit directly from above, the auxin will diffuse down the stem and stimulate growth equally on all sides.
- If the plant is illuminated from the side, the auxin will be moved from the light to the dark side, where it stimulates cell division and elongation.

- This causes the stem to grow faster on the dark side, causing the plant to bend towards the light. So an unequal distribution of hormone causes unequal growth.

Auxin is also responsible for a gravitropism. Again, an unequal distribution of auxin causes roots to grow downwards and shoots to grow upwards.

Auxin accumulates in the lower side of a germinating seedling, where it inhibits (slows down) root growth, but stimulates shoot growth. This results in the root growing downwards and the shoot growing upwards.

Uses of plant hormones

Revised

Because plant growth and development are controlled by hormones, we can use artificial plant hormones to our advantage in a variety of ways. We can control the rate and type of growth, timing of flowering and the ripening of fruit. You need to be aware of the following two applications:

1 Part of the stem is cut from the plant.

1   As selective **weedkillers** — ones that kill some plants but not others. Synthetic (man-made) auxins can be applied to a plant. Being a different shape from natural auxins, they cannot be broken down by the plant's enzymes, so they stimulate uncontrolled growth, causing over growth of some parts of the plant, but under-development in others, resulting in the plant's death

2 The leaves are removed from the bottom of the cutting.

When sprayed, many of these weedkillers are absorbed through the leaves rather than the roots. This means that the greater the surface area of the leaf, the more weedkiller is absorbed. In this way we can get rid of broad-leaved plants such as dandelions on a lawn. They will absorb a lethal dose before the narrow-leaved plants such as grass.

3 The cut end is dipped in a hormone rooting powder.

2   As **rooting hormones**. If you have a cut stem that you want to develop into a whole plant, dipping the cut end in rooting powder will stimulate root growth. Taking cuttings is an example of simple **cloning**, and as such is also an example of **asexual reproduction**.

4 The cutting is planted into compost.

Tip: The terms synthetic, man-made and artificial all mean the same thing.

↑ How to take a shoot cutting

Check your understanding

Tested

15 Why are plant responses slower than animal responses?   *(1 mark)*

16 Explain how a seedling on a windowsill will grow towards the light.
(3 marks)

17 If a particular weedkiller is toxic to all plants, how can spraying it on a lawn kill the dandelions but not the grass?   *(2 marks)*

18 When seeds start to germinate, why is it more important to respond to gravity than to light?   *(1 mark)*

Answers online   Test yourself online

Online

Testing and trialling medical drugs

Scientists the world over are constantly looking for new drugs. Before they can be granted a licence to be sold, medical drugs have to undergo a long series of tests to make sure that they are **effective** and **safe** to use. We also have to know what the **safe doses** are.

How drugs are developed

Revised

1 Many natural substances are tested, or **screened**, to see if they have any useful properties. Plants and other organisms are used throughout the world by native peoples as traditional medicines or poisons. These can be tested to see if they contain substances that may be useful — for example, as a new painkiller or antibiotic. Another approach is to start with existing drugs and modify them in the laboratory to see if they can be made more effective.

2 **Laboratory tests are carried out** on human **cell** and **tissue** samples. Sometimes this gives good results, but it cannot show the effects on the whole body. For instance, you cannot test the effectiveness of a sleeping pill on a sample of tissue.

3 The drug is tested for toxicity and effectiveness on **live animals**. This gives more information because it uses a whole organism, but it usually has to be a mammal because humans are mammals.

4 In **clinical trials** the drugs are tested on healthy human volunteers and/or patients with the relevant condition to see if they are effective and if there are any side effects. The trials start with a low dose, and then further tests are done to find the **optimum dose**.

Placebo effect and double blind trials

Revised

If a person thinks that they have been given an effective drug, they may feel better even if they have been given nothing more than an injection of salt water or a chalk tablet. This is the **placebo effect**, and it can make the results of clinical trials invalid. The solution is to do a **blind trial** in which patients do not know whether they have been given the active drug or not.

Another complication of the placebo effect is that patients can sometime tell whether they are being given the active drug or not just by the actions of the person giving them the drug. This problem led to the development of **double blind trials**, in which neither the patient nor the doctor/person giving the drug knows who has had the active drug until after the trial.

Thalidomide

Revised

In the 1950s, thalidomide was trialled as a sleeping pill or sedative. During the trials it was also found to be very effective against **morning sickness**, so it was also prescribed to pregnant women. The drug had not been tested on **pregnant women**.

Disastrously, the drug affected the **development of the fetus**, so that many children were born with very short and deformed arms or legs. Not surprisingly, the drug was banned in 1962 and more rigorous testing procedures, like those outlined above, were developed for new drugs.

Recently, thalidomide has made something of a comeback because it has been found to be very effective in the treatment of **leprosy** — a bacterial disease.

Check your understanding
Tested

19 Match the terms below to parts **a)–d)** in the paragraph. *(4 marks)*

clinical trials   **human tissue**   **side effects**   **live animals**

A promising new antibiotic has been found in a plant from the Amazon rainforest. The drug kills a wide variety of bacteria, but it must be tested to make sure that it has no _____**a)**_____. Firstly, the drug is tested on samples of _____**b)**_____. Then it is tested on whole organisms, which involves using _____**c)**_____ to check for toxicity. Finally, the drug is tested in _____**d)**_____.

20 Statins are drugs that lower blood cholesterol. You have been given a new statin to test. Outline the steps you would need to take in carrying out a clinical trial on a group of 100 people. Remember that it would not be ethical to deprive some people of a potentially life-saving drug just for the sake of a 'fair test'. *(4 marks)*

Answers online ——— **Test yourself online** ————————————————— Online

Legal and illegal drugs

There are basically two types of drug:

1 **Medical drugs** are medicines that are designed to treat a particular illness or condition. These drugs can still be used illegally, and can harm the body.

2 **Recreational drugs** are taken for their short-term effects. Many recreational drugs, such as cocaine, heroin and LSD, are illegal. However, the two drugs with the greatest effect on the nation's health — tobacco and alcohol — are legal if you are old enough.

The overall effect of the legal drugs (alcohol, tobacco and all the thousands of prescribed drugs) on the nation's health is much greater than that of the illegal drugs.

Recreational drugs
Revised

People take recreational drugs usually because they like their **short-term effects**, which include:

- relaxation
- stress relief — an escape from everyday worries
- stimulation — some drugs (e.g. amphetamines or 'speed') combat tiredness so that people can work longer or 'party all night'
- inspiration — some people claim to be more creative while under the influence of drugs
- pain relief — many multiple sclerosis sufferers claim that cannabis gives relief from muscular pain

However, there is the serious problem of **addiction**. Heroin and cocaine are highly addictive. Alcohol and cigarettes are also addictive, and users can suffer **withdrawal symptoms** if they try to give up.

Alcohol is legal but dangerous. Alcohol reduces the activity of the nervous system, leading to:

- slower brain function
- **slower reactions**, and **impaired coordination** and **judgement**, which is why it is illegal to drink and drive
- loss of **self-control**, so people do things they would not do when sober, such as saying 'yes' to sex

Excess alcohol in one session — binge drinking — can cause **dehydration** (the main cause of hangovers), **unconsciousness** and even a **coma**.

Long-term alcohol abuse can result in liver damage (**cirrhosis**) and **brain damage**.

Because so many people drink alcohol, the cost of dealing with alcohol-related health problems (millions of working days lost), accidents caused by drink-drivers, and the **social costs** to families can be very high.

Tobacco is legal in the UK for over-18s, while cannabis is illegal. Some people want to make cannabis legal, but this is controversial.

What is cannabis?

Cannabis is a plant, the leaves of which are dried and smoked, producing a feeling of relaxation and mellowness. Many people think it is relatively harmless — but it is still **illegal**.

Objections to cannabis include the following:

● Many studies have linked cannabis use with an increased incidence of **mental illness**, such as **schizophrenia** — this could be due to chemicals in cannabis that affect the brain.

● People say cannabis is a 'gateway drug' to harder, more addictive drugs — there is some evidence that most users of hard drugs tried cannabis first.

How do athletes cheat?

There is a variety of drugs that athletes can take to enhance their performance in some way. These are banned by the various sporting authorities. **Anabolic steroids** are drugs that mimic the male hormone testosterone. They are taken by athletes who want to improve their muscle size and power, but they can have a variety of side effects including, in the long term, heart failure.

Check your understanding

21 Imagine there is a hormone, produced by the brain, that gives you a feeling of wellbeing ('happiness'). It is an important part of homeostasis to keep the levels of this hormone more-or-less constant. If there is too little in the blood, the brain will make some more. If there is too much, the brain will stop production for a while.

A drug becomes available that has the same effect as this hormone.

a) What will the drug-user feel in the short term? *(1 mark)*

b) Explain how the brain will respond to the high levels of this drug. *(1 mark)*

c) What will the drug-user feel after the effects of the drug have worn off? *(1 mark)*

d) What does addiction mean? *(1 mark)*

e) Use this information to explain how people can become addicted to a drug. *(1 mark)*

22 List two direct ways and two indirect ways that alcohol abuse leads to health and social problems. *(4 marks)*

23 Evaluate the statement 'Cannabis should be legalised'. *(2 marks)*

Adaptations

All organisms have the same basic aim: to eat and not be eaten until they have a chance to reproduce. Usually, this is difficult: most resources are in short supply and so there is **competition** between different species, and between individuals of the same species.

Plants usually compete for:

● light

● water

● soil nutrients

● carbon dioxide

Animals might compete for:

● food

● territory

● mates

● nesting sites

All organisms need to be adapted to suit their particular environment. Often, an extreme environment can result in extreme adaptations.

Living in extreme conditions   Revised ☐

Some bacteria can survive in extreme conditions that would kill most organisms instantly. Examples include:

● hot springs where the water is nearly at boiling point

● lakes where the water is very acidic, alkaline, salty or polluted

● deep in the ocean where the pressure is very high and water can become superheated (well above 100°C) around underground thermal vents

These bacteria are called **extremophiles** ('extreme loving') and have enzymes that are not damaged, or **denatured**, as easily as those of 'normal' organisms.

Polar animals   Revised ☐

Polar animals are adapted to reduce heat loss.

● Many polar animals have a compact, rounded shape to keep their **surface area** as **small as possible**. This minimises heat loss.

● They also have a thick layer of **blubber** — tissue rich in fat that **insulates** them from the cold. This fat can also act as a **food reserve**.

● Thick **fur** traps a layer of **insulating air**.

● Animals such as the polar bear and arctic fox have white fur so that they are **camouflaged** — making it easier for them to hunt in snow.

↑ **The walrus is a large animal with a thick layer of blubber, so that it can swim in the coldest arctic waters without losing too much heat**

Desert plants

Revised

Desert plants are adapted to reduce water loss. Most plants need a regular supply of water, but the cactus can survive in the desert by **absorbing** as much water as it can when available, and **losing as little as possible**.

● The leaves of most plants have a large surface area to trap light, but they also lose a lot of water. In the cactus the leaves have been reduced to spines. This **reduces surface area** for water loss and provides useful protection.

● Cacti have a **thick, fleshy stem** that can **store water**, and a **thick, waxy outer layer** so that very little water is lost by evaporation.

● Some have shallow but **extensive roots** to absorb water when it rains. Others have **deep roots** to reach underground water.

→ Survival in the desert is all about minimising water loss

Avoiding predators is vital

Revised

There are many different ways to avoid being eaten.

● Some plants and animals have **weapons** and **armour**. Plants such as roses, cacti and hawthorn have spines or **thorns**, and animals such as hedgehogs and porcupines have **spines**.

● Some are **camouflaged** so that predators do not notice them. Stick insects look like the plants they feed on, and are very slow so that predators do not notice their movement.

● Some animals produce **toxins** (poisons), and often use **warning colours** to advertise the fact. Wasps and poison arrow frogs are good examples.

> **examiner tip**
>
> You may well be asked a question about a plant or animal you have not studied. Don't panic — you should be able to look at the information given and figure out how that plant or animal is adapted to survive.
>
> The relationship between surface area and heat loss is a particular favourite with examiners, so make sure you know the basic idea.

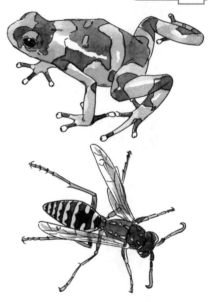

↑ 'Go away, I'm poisonous' — a poison arrow frog and a wasp

Check your understanding

Tested

24 State two ways in which a prey animal may be adapted to deter predators. *(2 marks)*

25 A fieldwork survey measured the distribution of moss growing in different parts of the school grounds. More moss was found on grass underneath a tree compared with the grass on the football field.

Suggest two environmental factors that could be responsible. *(2 marks)*

26 Describe and explain three ways in which a camel is adapted to life in the desert. *(3 marks)*

Answers online   **Test yourself online**

Online

Environmental change

Changes in the environment can be **natural** or they can be **man-made**. They could be due to:

● **living** factors such as predation and food supply

● **non-living** factors such as temperature, rainfall, oxygen levels and pH.

Either way, climate change can have a drastic effect on the distribution of organisms.

Measuring environmental change

Environmental change can be measured directly:

● **Oxygen levels** can be measured using an **oxygen meter** — a sensitive **probe**.

● **Temperature** can be measured using accurate maximum–minimum **thermometers**, which record the highest and lowest temperature on any particular day. The UK has reliable temperature records dating back to 1640.

● **Rainfall** is measured using a rain gauge or **pluviometer**, which is basically a measuring cylinder that catches the rain.

We can also measure environmental change by its **effect on organisms**. Certain species of plant and animal are great **indicators** of environmental change.

Lichens are very sensitive to air pollution. If lichens begin to disappear, this strongly suggests a decline in air quality. In the UK the air quality has got better over the last 50 years.

↓ **Several species of invertebrate animal are reliable indicators of water quality**

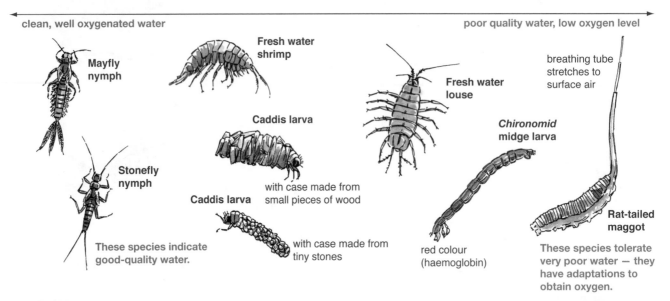

clean, well oxygenated water → poor quality water, low oxygen level

Mayfly nymph

Fresh water shrimp

Stonefly nymph

Caddis larva

Caddis larva
with case made from tiny stones

with case made from small pieces of wood

These species indicate good-quality water.

Fresh water louse

Chironomid midge larva

red colour (haemoglobin)

breathing tube stretches to surface air

Rat-tailed maggot

These species tolerate very poor water — they have adaptations to obtain oxygen.

Global warming

The Earth's atmosphere has been getting warmer in the past 200 years. Scientists think that this is due to an increased **greenhouse effect**.

As in a greenhouse, the Earth's atmosphere allows **sunlight energy in**, but does not let all of it escape, so it gets hotter.

1   Temperatures on Earth result from a **balance** in **energy received** and **energy lost**.

2   The energy we get from the Sun is not changing, but we are **losing less**.

3   This is because we are **changing** the **composition** of our **atmosphere**. Gases such as **carbon dioxide** and **methane** act together to prevent heat from escaping.

Power stations, industry, cars and aircraft engines all **burn fossil fuels** to produce millions of tonnes of carbon dioxide every year. While there is a massive amount of **combustion** going on, **photosynthesis still absorbs carbon dioxide** from the atmosphere.

Cutting down forests (**deforestation**) is another problem. A lot of carbon is locked up in wood, so when we cut down and burn trees:

● more carbon dioxide is released

● there are fewer trees to remove carbon dioxide from the atmosphere

Effects of global warming
Revised

The Earth's climate is a **complex system**. This means that its effect is impossible to predict exactly. However, here are some possible effects of global warming:

● **Glaciers** and **ice caps** on Greenland and Antarctica will **melt** and drain into the sea, making **sea levels rise**.

● Higher temperatures make water **expand**, which scientists think may have more effect than melting ice on rising sea levels.

● Rising sea levels will flood low-lying areas with high populations, such as large areas of the Netherlands.

● The extra fresh water may **disrupt ocean currents**. This could be a disaster for Britain, which is currently kept warmer by Atlantic currents.

● More **severe weather** could result — for example more **hurricanes**.

● Higher temperatures could mean more evaporation, more clouds and a change in rainfall patterns.

● All over the world, ecosystems depend on the climate. Changes could disrupt plant growth and patterns of agriculture.

How can we be sure?
Revised

The evidence says that global warming is occurring, that ice caps are melting, and that all this is due to human activity.

● There is a **correlation** between increased greenhouse gases and global warming.

● A correlation may be a **coincidence**, but the rising temperature agrees with **climate prediction models** carried out by **different scientists** worldwide, making the conclusion more **plausible**.

The models also show that the effect of a rise of just a few degrees will be damaging. We may not feel the difference in our lifetime — but changes are clearly happening, so we must do something for future generations.

↑ **The Maldives are a group of islands in the Indian Ocean. As much as 80% of the Maldives are less than 1 m above sea level. The most gloomy predictions say the Maldives could sink beneath the waves within the next few decades**

Check your understanding
Tested

27 List two factors contributing to increased carbon dioxide levels in the atmosphere. *(2 marks)*

28 Explain the effect of deforestation on carbon dioxide concentrations in the atmosphere. *(2 marks)*

29 Use the example of carbon dioxide and global warming to explain the difference between a correlation and a cause. *(2 marks)*

Answers online   **Test yourself online**
Online

Energy in biomass

- In **photosynthesis**, plants and algae use light energy to turn carbon dioxide and water into glucose.
- The plant converts glucose into other sugars, starch, cellulose, lipids and proteins.
- In this way the plant increases its **biomass** — the **weight of living material**.
- Some of the **energy** in the biomass in the plant is **transferred** when the plant is eaten, and so works its way up the **food chain**.
- These **energy transfers** are **not efficient** — most of the energy and biomass is lost.

examiner tip

Light energy, sunlight, solar radiation — they all mean the same thing.

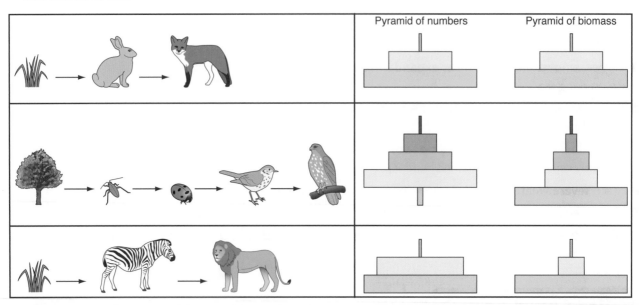

↑ Pyramids of numbers and pyramids of biomass for three simple food chains. The arrows show the direction of energy flow

Interpreting food chains
Revised

The stages in a food chain are called **trophic levels**. The organisms on the first trophic level are usually plants, and are called **producers** because they produce the food.

After the producers come the **consumers**. **Primary consumers** are animals that eat the producers. **Secondary consumers** eat the primary consumers, and so on. Primary consumers are usually herbivores such as rabbits, cows or deer, while secondary and tertiary (third level) consumers are usually carnivores.

Interpreting pyramids of biomass
Revised

A **pyramid of numbers** shows the size of the different populations at each stage in the food chain, but it does not account for the huge variation in size between different species, such as between grass and oak trees.

- To construct a **pyramid of biomass**, calculate the biomass of each species in the food chain.

- The area of the rectangle representing each trophic level is proportional to the total biomass at that level. A wide rectangle equals more biomass than narrower rectangle.
- Notice that the biomass at each stage in a food chain is less than it was at the previous stage.

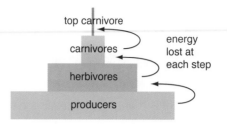

↑ Not all the energy contained in biomass is transferred at each level in a food chain

Where does all the energy go?

Revised

The diagram on the right shows that only a small proportion of energy in the biomass is transferred to the next trophic level. The **efficiency of energy transfer** is usually less than 10%.

1 Not all of the energy from the Sun that reaches plants can be used for photosynthesis. Plants cannot use green light, for example, and some light passes through the leaves.

2 Of the biomass that is made in photosynthesis, some is used by the plant to **respire**. Respiration supplies all the energy needs for the plant to maintain itself and grow.

3 When an animal eats the plant, a lot of the plant cannot be digested. The only energy available to the animal is in the food molecules that it **absorbs**. The energy in undigested food (such as cellulose from cell walls) passes right through the gut and is lost in faeces.

4 Animals respire too. Energy released from food in respiration is used by the animal to move, and for chemical reactions in living processes. A large proportion of this energy is lost to the environment as heat, especially in **mammals and birds**, which are warm blooded. They need to respire quickly to replace lost heat.

These **energy losses** are why most food chains have only two, three or four links — the energy runs out. Large predators at the top of a food chain — known as apex predators — are usually rare because there is only enough energy to support a small population.

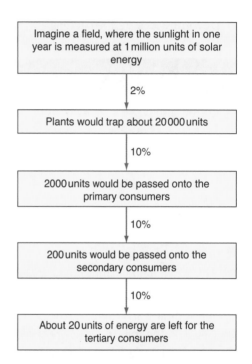

↑ How to lose a million units of energy

Check your understanding

Tested

30 Define the following: **a)** producer; **b)** primary consumer; **c)** trophic level. *(3 marks)*

31 Construct two separate food chains from the organisms listed below. *(2 marks)*

| antelope | blackbird | cat | detritus |
| grass | cheetah | earthworm | |

32 Algae and certain types of bacteria are not classed as plants but they can be classed as producers. What can they do that makes them producers? *(1 mark)*

33 Explain how the efficiency of energy transfer accounts for the shape of pyramids of biomass. *(2 marks)*

Answers online   Test yourself online   Online

Decay processes

What happens to waste materials?

The simple answer is that these materials are **recycled**, mainly due to the action of microorganisms — **bacteria** and **fungi**.

Waste material from animals and plants includes:

- dead leaves and other plant material
- dead bodies
- faeces
- urine

All this material contains lots of **carbon** locked up inside protein, carbohydrate and lipid molecules. These molecules are broken down when the material **decomposes**, or rots.

- **Detritus feeders** (or **detritivores**), for example woodlice and earthworms, often start the process of decay by breaking plant tissue into much smaller pieces.
- **Decomposers** are the bacteria and fungi that make things rot.

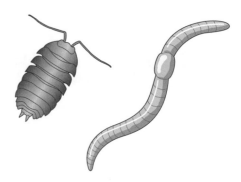

↑ **Earthworms and woodlice are detritivores; they feed on rotting vegetation**

Rotting is bacteria and fungi feeding

When the spores of bacteria and fungi land on anything they can feed on, such as dead leaves, food leftovers or a dead animal, they start to **digest** it (break it down) by releasing **enzymes**. The enzymes make the 'food' simple and soluble, so the decomposers can absorb the soluble products straight into their cells.

The ideal conditions for rotting are:

- warm
- moist
- aerobic (i.e. oxygen is available)

Oxygen is needed because the microorganisms respire. Carbon dioxide is released, just like it is from animals and plants.

Mineral ions and any nutrients the bacteria or fungi do not need are released into the soil. This is where things come full circle: many of these by-products, such as **nitrate** and **phosphate**, are essential for plant growth.

In a **stable** ecosystem, all of these processes will be **balanced**. The **producers**, **consumers** and **decomposers** are all **interdependent**, and so materials are constantly recycled.

↑ **Soil is a mixture of rock fragments (clay, sand, pebbles), water, air, organisms and humus. Humus is dead material that is being broken down by bacteria and fungi. It releases vital minerals such as nitrate and phosphate, which the plants absorb through their roots**

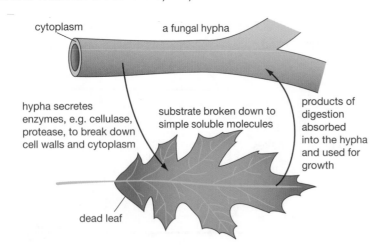

cytoplasm

a fungal hypha

hypha secretes enzymes, e.g. cellulase, protease, to break down cell walls and cytoplasm

substrate broken down to simple soluble molecules

products of digestion absorbed into the hypha and used for growth

dead leaf

← **Fungi feed by extracellular digestion**

The carbon cycle

The carbon in our bodies was once part of the food we ate. Before that it was part of a green plant, and before that it was part of carbon dioxide in the atmosphere. When our bodies die, carbon is released into the atmosphere as carbon dioxide when decomposers respire.

Photosynthesis is the process that turns carbon dioxide back into sugars again.

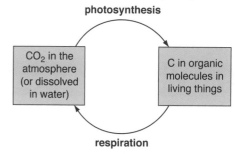

Respiration is the process that turns sugars into carbon dioxide.

→ At its most basic level, the carbon cycle is a flow between carbon in carbon dioxide in the atmosphere (or dissolved in water) and the carbon in organisms

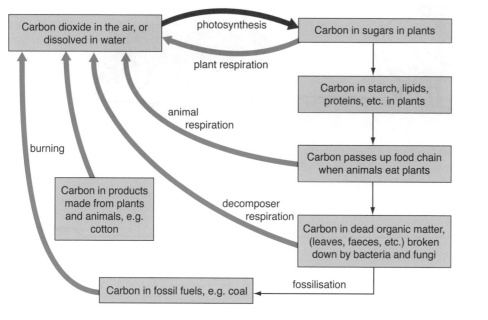

← A more complete carbon cycle that includes plants, animals and decomposers

1 Photosynthesis is the **only process** that takes carbon dioxide out of the atmosphere. The other processes, **respiration** and **burning**, put carbon dioxide back into the atmosphere.

2 Human activities are changing the balance. There is more burning and less photosynthesis as areas of forest are cleared. Carbon dioxide should be released into the atmosphere at the same rate as it is absorbed by plants for photosynthesis. However, human activity puts much more carbon dioxide into the air than can be removed by the plants — adding to **global warming**.

In the end, all of the sunlight energy that was originally captured by plants in photosynthesis is **released** into the **atmosphere** via the processes of respiration or combustion.

Check your understanding
Tested

34 a) Explain what happens when something rots. *(3 marks)*

   b) Explain why the process of decay is essential to plants. *(2 marks)*

   c) Name two types of organism that cause decay. *(2 marks)*

35 Match each of the following words to the statements below.

   respiration decay photosynthesis fossilisation combustion

   a) Removes CO_2 from the atmosphere. *(1 mark)*

   b) Turns dead plant and animal remains into coal, gas or oil. *(1 mark)*

   c) Living things releasing CO_2 into the atmosphere. *(1 mark)*

   d) Microbes breaking down dead organic matter. *(1 mark)*

   e) Releases CO_2 into the atmosphere when fossil fuels are burned. *(1 mark)*

Answers online — **Test yourself online** Online

Why organisms are different

It is obvious that all people are different — even identical twins will have their differences. **Variation** is to be found in all other species too.

How do organisms vary?

Revised

Consider your year group at school. You can see variation in countless different ways: height, weight, eye colour, shoe size, blood group and intelligence, to name but a few.

↑ **Dogs and sheep look very different from each other because they are different species — but despite what you might think, the sheep all vary too**

Cause of variation

Revised

There are two causes of variation:

1 Your **genes**:
- All living things have **characteristics** that are **similar** to those of their **parents**.
- This is because an organism's characteristics are **controlled by genes**.
- Genes are **instructions** that control the way an **organism grows and develops**.
- We inherit a **random mixture** of genes from both parents, which is why we are different, even from our brothers and sisters.

2 Your **environment**:
- This refers to the conditions in which you have lived and grown up.
- It includes the food you have eaten, the experiences you have had and the pathogens you have been exposed to.

So are you the way you are because of your genes, or is it your environment? The answer is, a complicated mixture of both. People often call this the **nature versus nurture** argument.

Some characteristics are determined by **genes only**. Examples include eye and hair colour (yes, you can dye your hair, but that's cheating), blood group and inherited diseases such as cystic fibrosis.

Other factors, such as height, are controlled by both — you will inherit the genes to grow to a certain height, but you will not do so without a good diet.

Genes, chromosomes and DNA

Revised

When you first looked down a microscope at some cells, perhaps from your cheek, you will have seen the **nucleus** as just a dot. But what really goes on in the nucleus is the key to life itself.

1   The nucleus contains a lot of DNA. Usually, it is all spread out so you cannot see any detail.

2   If a cell is going to divide, it will roll its DNA up into X-shaped structures called **chromosomes**.

3   Most human cells have 23 **pairs of chromosomes**. There are two number ones, two number twos, etc. We cannot draw them all here.

4   A chromosome is one long, tightly coiled DNA molecule. The **genes** are regions of DNA, like words on a long piece of tape. There are hundreds or even thousands of genes in a single chromosome.

Different species have different numbers of chromosomes. Humans have 46 but cats, for example, have 38. Don't make the mistake of stating that all species have 46 chromosomes.

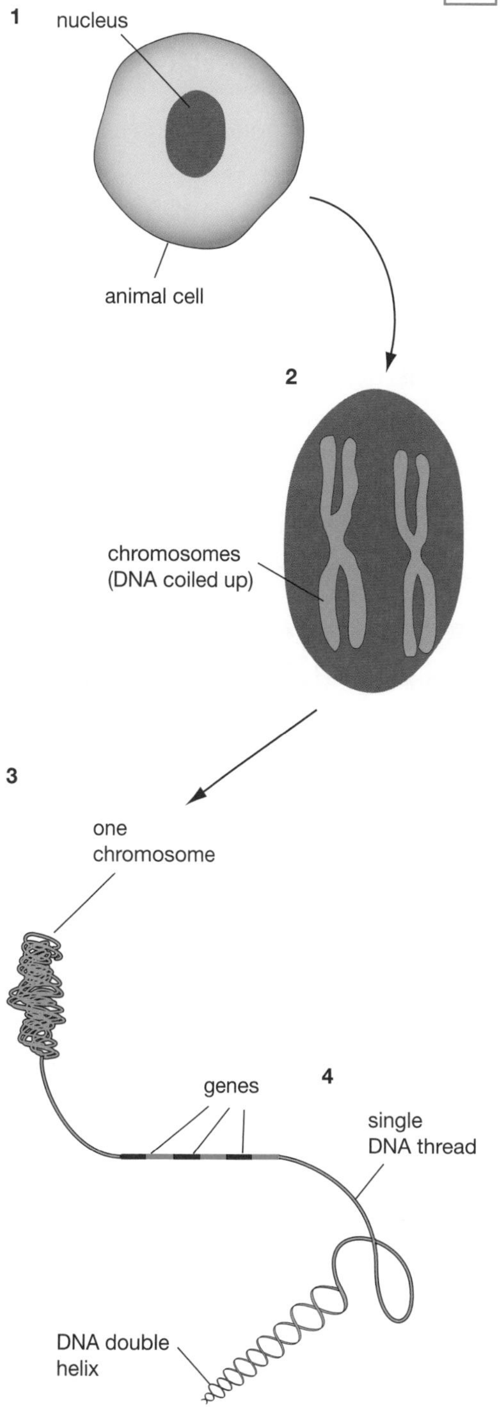

1 nucleus

animal cell

2

chromosomes (DNA coiled up)

3

one chromosome

4

genes

single DNA thread

DNA double helix

examiner tip

You could compare the contents of the nucleus to a book. Each chromosome is a chapter, and each gene is a different word.

Check your understanding

Tested

36 Identical twins have exactly the same genes, so you would expect that any differences would be due to their environment.

   a) Suggest a reason why studies on identical twins show less variation in characteristics controlled by the environment compared with the variation observed in other schoolchildren.

(2 marks)

   b) Suggest a possible reason why studies of identical twins who were raised apart are often seen to be not valid.   *(1 mark)*

Answers online   Test yourself online

Online

Reproduction and cloning

Asexual reproduction gives clones

- **Asexual** means non-sexual.
- Asexual reproduction involves just **one parent**.
- There is no mixing of genetic information, so all cells made are genetically identical to the parent — they are **clones**.
- In asexual reproduction, a whole new organism is made.
- Many different species reproduce asexually — bacteria, yeasts, many plants and some animals (such as greenfly).

→ **Asexual reproduction in a strawberry plant**

Sexual reproduction gives variety

- **Sexual** reproduction involves **two parents**.
- There is a **mixing** of **genetic information** so that each new individual is **unique**.
- In sexual reproduction, the parents produce **sex cells**, or **gametes**.
- The male gametes are **sperm**, the female gametes are **eggs** (**ova**).

Cloning

Identical twins result from **one fertilised egg splitting into two**, each of which develops into a baby. This is natural cloning.

Cloning plants is relatively easy.

1. **Taking cuttings**: gardeners or farmers can simply take a cutting from a plant they want to clone, put it into the right conditions, and it will grow into a copy of the parent plant.

2. **Tissue culture**: a small piece of plant tissue — perhaps just a few cells — is put into a growth medium. The cells multiply, and are then separated. Each cell in its own growth medium multiplies into a clump of cells that grows into a new plant.

Whole animals can be cloned in two ways. Farm animals, such as sheep and cattle, do not reproduce asexually, but there are ways to produce identical copies of individual animals.

What about cloning humans?

1. Can we? — the **science** question.

2. Should we? — the **ethics** question.

The answer to question 1 is not yet — the process is just the same as cloning a sheep but no human has been cloned yet because the process is not reliable. The answer to question 2 depends on an individual's opinions, which may be influenced by **religious** or **ethical** beliefs.

Embryo transfer

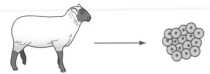

1. An embryo (clump of developing cells) is removed from a pregnant animal.

2. The embryo is split into a number of smaller clumps of cells.

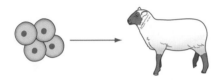

3. Each new embryo is inserted into the uterus of another host mother.

4. Some of the host mothers become pregant and give birth to offspring. The offspring are clones of the original embryo, not the mother from which it was taken.

Adult cell cloning

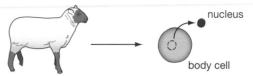

1. Take a cell (e.g. from the skin) from the animal to be cloned, and remove its **nucleus**. Remember that a **full set of genes** is presen in the nucleus of **every cell**.

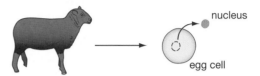

2. Take an **unfertilised egg** from the same species and remove its **nucleus**.

3. Put the nucleus from the animal to be cloned into the unfertilised egg.

4. The **fused** egg is stimulated with a **mild electric shock** so that it starts to develop into an **embryo**.

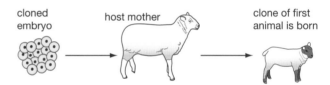

5. Implant the embryo into a uterus. The embryo is a clone of the adult from which the single cell was taken.

↑ Embryo transfer and adult cell cloning

There are several ways of cloning human tissue for treatment. **Tissue culture** could be used to grow cloned replacement tissue, for example skin for burns patients. Tissue could also be grown from an embryo cell that had its nucleus replaced with one from the patient. However, cloning cells from embryos kills the embryo. Alternative methods are being researched, such as **fusion cell cloning**. A single adult cell (for example, from skin) is **fused** with cytoplasm from an egg cell. The fused cell (a clone of the original adult cell) can divide into many different types of cell.

Check your understanding
Tested

37 Describe the difference between adult cell cloning and embryo transfer techniques. (2 marks)

38 When a gardener takes a cutting, each cutting grows into a new plant that looks identical to the original plant. Explain. (1 mark)

Answers online —— Test yourself online
Online

Genetic engineering

Using **genetic engineering**, scientists can add to, remove or alter an organism's genes. There are exciting possibilities (for example, genetically engineered bacteria can make insulin for diabetes) — but there might be problems too.

Changing characteristics by gene transfer   Revised

Remember that different genes control the development of different characteristics. The process of genetic engineering is as follows:

1   Find a gene for a useful characteristic in one organism.

2   **Cut it out** of the chromosome using an **enzyme**.

3   Cut open the DNA of the target organism.

4   Insert the cut-out gene into the organism's DNA.

5   Put the new **modified** DNA back into a cell.

6   As the cell grows and multiplies, all the new cells have copies of the modified DNA, and so have the desired characteristic.

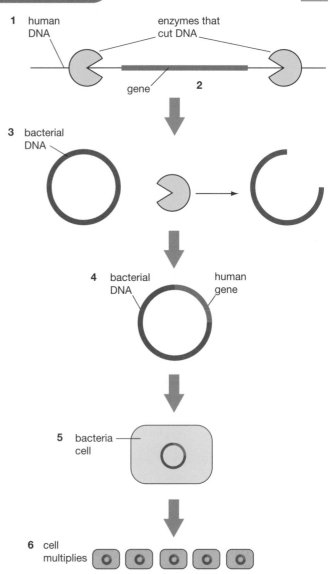

Producing new varieties of crops   Revised

Genes from one plant can be transferred into another species of plant. This can produce a new variety with **increased yield**, or **better nutritional value**.

Genes from bacteria can also be transferred into plants. For example, a certain bacterium produces a substance that kills insects that destroy crop plants. Scientists cut out the **insecticide gene** and transfer it into crop plants such as maize and cotton. The **genetically modified (GM)** crops now produce their own insecticide, and resist insect attack.

Modifying animals and plants

Revised

Genes can also be transferred to animal **embryos**. A solution containing genes that carry a desired characteristic (such as disease resistance) is **injected** into the embryo cells. The embryo develops with the desired characteristic.

In plants, tiny metal particles coated with genes can be fired into plant cells. The full plants grown from these plant cells have the new characteristic.

Benefits for food production

Revised

As well as GM crops with better yield, nutritional value or insect resistance, we could produce:

● crops that are **resistant** to **disease** or **herbicides**

● salt-tolerant crops that can grow in salty areas that cannot be farmed at the moment

● crops that grow more quickly (by putting the growth genes from fast-growing species into slower-growing species)

● sheep that grow up to make useful products in their milk

Problems

Revised

● Some people are concerned about the effect of eating GM crops on **human health**.

● Some people are worried that transplanted genes will **spread** into **wild plant** populations by cross-pollination. For example, there is a risk that common plants could become 'superweeds'.

● If other (harmless) insects feed on a GM crop that kills insect pests, biodiversity could be reduced.

● If farmers know they can spray weedkiller over a whole field and not harm the crop, will they kill all the other plants, including wild flowers?

examiner tip

If a question asks you to *evaluate* a particular point, you need to look at both sides. For example — state the advantages and disadvantages of GM crops.

Check your understanding

Tested

39 Give an advantage to farmers of growing herbicide-resistant crops.

(1 mark)

40 A variety of soya bean has been genetically modified to be resistant to herbicides. Some people are worried that GM crops like this may lead to a reduction in biodiversity. Suggest an explanation for this.

(2 marks)

Answers online   Test yourself online   Online

Evolution

Natural selection
Revised

The idea of evolution had been around for centuries, but it was **Charles Darwin** who first put forward a good idea about how it could happen. Darwin's theory was based on the following observations:

1 There are more organisms born **than can possibly survive**.

2 So there will be **competition** for food, mates, etc.

3 There is usually **variation** within a population due to individuals' genes (see page 59).

4 In the struggle to survive, some will be born with an advantage (see page 22). Their **genes** make them **better adapted** to their environment.

5 There is more chance that a better-adapted individual will **survive** and **pass on its genes** to the next generation. The offspring that inherit these genes will have also an advantage, and so it goes on. This is **natural selection**.

6 Organisms that reproduce quickly, such as bacteria, will have more mutations in a given time. This means they can **adapt** to changing conditions more rapidly than a slow-reproducing species.

Evolution starts with mutations
Revised

A **mutation** is a change in an organism's **DNA**. There are three possible outcomes of a mutation:

1 It may be harmful, killing the organism or giving it a disadvantage in terms of survival.

2 It may have no effect at all.

3 Very occasionally it may be useful, for example giving new coloration, better eyesight or a stronger immune system. In this case the mutation will give the individual a **survival advantage**.

Darwin's theories were controversial
Revised

Darwin came up with his idea for evolution in his twenties, but he knew that it would cause controversy, so he gathered evidence to support his theory for decades, and did not publish until he was about 50 years old.

- Darwin's theory went against common religious beliefs, which were based on the Bible. This said that God made all the plants and animals that live on Earth.

- There is now much evidence to support Darwin's theory. The fossil record grows more complete all the time. There are new techniques that allow us to compare the DNA and proteins of different organisms. These provide even stronger evidence for evolution, and help us to build evolutionary family trees that **model** the most likely pattern of evolution.

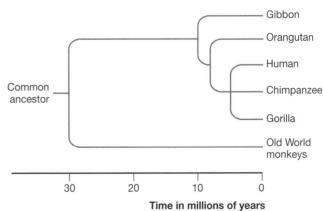

↑ **Evolutionary tree of humans and great apes**

- Darwin could not explain **how** characteristics were varied or passed on. This was one of the biggest criticisms from other scientists at the time.
- Breakthroughs that supported Darwin's ideas — such as DNA, genes, chromosomes and cell division — were not discovered until 50–100 years after Darwin published his ideas.

Other theories of evolution

Revised

Scientists may produce different **hypotheses** to explain similar observations. In 1809, the year Darwin was born, a Frenchman called Lamarck put forward an idea called the '**inheritance of acquired characteristics**'.

- Lamarck said that **useful changes** in a species were **acquired** during the organism's lifetime and then inherited by its offspring.
- We now know that the environment can change characteristics (eating different food; going to the gym), but these changes **cannot be inherited**.

→ *Lamarck's explanation for the heron's long legs. It is a big help to have long legs, because the heron can hunt in deeper water and catch more fish. So over the course of the heron's lifetime, the legs became longer and these characteristics were passed on to the offspring*

Classifying organisms

Revised

There are millions of different species known to science, and many more yet to be discovered. Organisms are classified according to the **features they have in common**.

Organisms are classified into large groups called **kingdoms**, the main ones being the **animals**, the **plants**, the **fungi** and the **bacteria**.

Within the animal kingdom are the **vertebrates** and the **invertebrates**. The vertebrates include the **mammals**, **fish** and **reptiles**. The mammals include the rodents, primates and whales, and so on. As the groups get smaller, the similarities between the organisms become more obvious.

The big idea in classification is to build a giant catalogue of life on earth, showing the story of what evolved from what. This may never be possible because the extinct species greatly outnumber the living ones, and many of them did not leave fossils behind.

Check your understanding

Tested

41 Can you think of any evidence that disproves Lamarck's idea? *(1 mark)*

42 What is the key difference between Darwin's and Lamarck's theories? *(2 marks)*

43 Explain how a population of rats could evolve resistance to a new type of rat poison. *(4 marks)*

Answers online   **Test yourself online**

Online

Cells and cell structure

All living things are made up of cells
Revised

- Cells are tiny compartments of living tissue, too small to see without a microscope.
- Many living things, for example bacteria, have **just one cell**.
- Cells cannot be large, so large organisms like humans are **multicellular** — made from millions and millions of cells.
- Being multicellular allows **different** cells to have **different** jobs.
- You need to know about four different types of cell: **animal**, **plant**, **yeast** and **bacterial**.
- These four types represent major **kingdoms** in the living world.
- Animal, plant and yeast cells are **larger** and **more organised** than bacterial cells.

Human cells are like other animal cells
Revised

Animal cells are the only type of cell that does not have a cell wall. They just have a membrane, which acts as a barrier between the cell and its environment.

- The nucleus contains the **DNA**. DNA exists in long molecules that coil up into **chromosomes**. Along the length of each DNA molecule are specific regions called **genes**.
- Each gene contains the instructions for the production of a particular **protein**. Many of these proteins are **enzymes**.
- Ribosomes are the site of **protein synthesis**, i.e. where proteins are made.
- Mitochondria are the site of **aerobic respiration**, a process that releases the chemical energy contained in molecules such as glucose and fats (lipids).

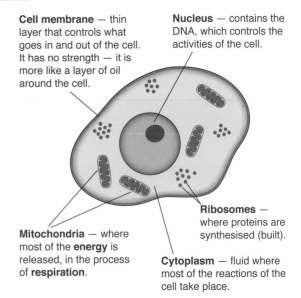

Cell membrane — thin layer that controls what goes in and out of the cell. It has no strength — it is more like a layer of oil around the cell.

Nucleus — contains the DNA, which controls the activities of the cell.

Ribosomes — where proteins are synthesised (built).

Mitochondria — where most of the **energy** is released, in the process of **respiration**.

Cytoplasm — fluid where most of the reactions of the cell take place.

↑ The essential features on an animal cell

Plant cells are different
Revised

The cells of plants and algae always have a cell wall made from **cellulose**. Algae are not classed as plants, but they do have similar cells to plants and they also **photosynthesise**.

Vacuole — large space filled with fluid (**sap**). Helps to keep the cell turgid (inflated) like a bladder in a football.

Chloroplasts — contain chlorophyll, which is needed for **photosynthesis**.

Cell membrane

Cell wall — made from **cellulose**, a tough, 'stringy' material. Gives the cell shape and makes it rigid.

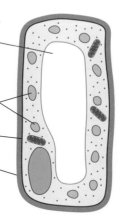

← The essential features of plant and algal cells

examiner tip

Plant cells have all of the things that animal cells have. Plant cells also **always have a cell wall**, and often a permanent **vacuole** and **chloroplasts**.

Bacteria
Revised

The first living things ever to evolve were similar to present-day bacteria. Compared with the other cell types on this page, bacteria are very small and simple.

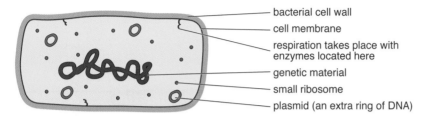

bacterial cell wall
cell membrane
respiration takes place with enzymes located here
genetic material
small ribosome
plasmid (an extra ring of DNA)

← A bacterial cell. There any many different species of bacteria, such as *E. coli*. All have the same basic features

Features of bacteria:

● They do not contain many of the internal structures found in other cells, such as a nucleus with a membrane, or mitochondria.

● All of the cell processes take place in the cytoplasm, rather than in a particular structure.

● Respiration takes place on folded parts of the membrane.

● Bacteria have ribosomes for protein synthesis, but smaller than the ones found in plant, animal and fungal cells.

● There is no true nucleus. Instead there is one **central chromosome** that contains all the genes essential for life. In addition, there are tiny rings of DNA called **plasmids** that contain extra genes. Plasmids are very useful in genetic engineering (see page 34).

Yeast
Revised

Yeast is a single-celled fungus. There are lots of multi-cellular fungi too, such as those that can be seen growing on rotting food. All fungal cells have a **cell wall**, but it is **not** made from cellulose.

All fungi feed by extracellular digestion — they make and secrete enzymes that digest the surrounding material.

Yeast are of particular interest because when they respire **anaerobically** (see page 56) they make **carbon dioxide**, which makes bread rise, and **alcohol**, which is popular in certain drinks.

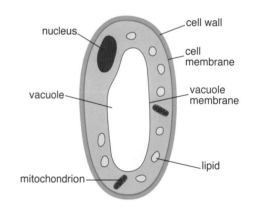

nucleus
cell wall
cell membrane
vacuole
vacuole membrane
lipid
mitochondrion

→ A yeast cell

Check your understanding
Tested

1 The following are components of cells:

   nucleus   chloroplasts   mitochondria   ribosomes

   Which one would you expect to be particularly abundant in:

   a) a muscle cell? *(1 mark)*

   b) a palisade cell in the upper surface of a leaf? *(1 mark)*

   c) a pancreatic cell that makes digestive enzymes? *(1 mark)*

2 Explain your answer to each part of question 2. *(3 marks)*

Answers online   Test yourself online   Online

Dissolved substances

- All living cells contain fluid — the **cytoplasm** — and most are surrounded by fluid too.
- Cells are constantly exchanging materials with their surroundings.
- This section is about how dissolved substances get in and out of cells.

How substances get in and out of cells
Revised

To get in or out of a cell, substances have to cross **cell membranes**. There are two basic ways of doing this: by **diffusion** or **osmosis**.

In a liquid or a gas, particles are continually moving around — this is called **kinetics**. When particles are concentrated in only part of a region, they will move around randomly until they are evenly spread throughout the whole region.

Diffusion is the net movement of particles from a region of high concentration to a region of lower concentration until they are evenly spread.

examiner tip

'Particles' is a general term for molecules or ions.

A good way to think of diffusion is simply 'stuff spreads out', but examiners will not be too keen on your writing that in the exam.

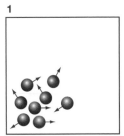
Concentrated in one area…

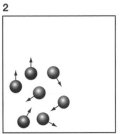
…molecules bounce around…

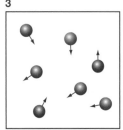
…until evenly spread

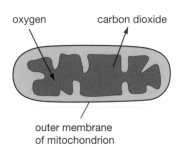

↑ Oxygen diffuses into mitochondria, where it is vital for aerobic respiration. Carbon dioxide, a waste product, diffuses out.

Importance of diffusion
Revised

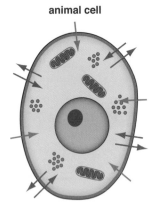
animal cell
- oxygen required for respiration

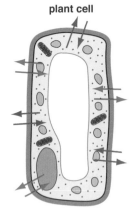
plant cell
- oxygen required for respiration
- oxygen produced by photosynthesis

Key:
→ oxygen
→ water

← The movement of oxygen and water into and out of animal and plant cells. Water passes in either direction by osmosis. Animal cells use oxygen; plant cells can make or use oxygen

Diffusion is the process by which gases and dissolved substances move from one place to another. This can be both into or out of an organism, and from place to place within an organism, such as in and out of cells.

Examples of diffusion:

- In lungs, oxygen diffuses from air into blood.
- Oxygen diffuses from blood into cells, where it is needed for **respiration**.

- Carbon dioxide diffuses from blood to the air in your lungs.
- Small molecules of digested food diffuse from your gut into your blood.

Concentration gradient and diffusion speed

Revised

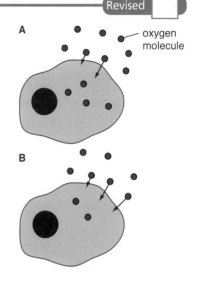

A

oxygen molecule

B

- The greater the difference in **concentration**, the faster the rate of diffusion.
- The idea of **maintaining a diffusion gradient** is a very important one in biology.
- For example, we **breathe** to make sure that there is always fresh air in our lungs, so there is more oxygen in the air in our lungs than in the blood. There is always a diffusion gradient, and so oxygen keeps passing into the blood.

→ Diffusion of oxygen into cell B takes place faster than into cell A because there is a bigger difference between the concentrations of oxygen inside and outside cell B

blood capillary in lung

alveolus

oxygen diffuses into red blood cells

carbon dioxide diffuses from the blood into the alveolus

thin layer of water surrounding alveolus

alveolus membrane has thin layer of flattened cells giving a short diffusion distance

capillary wall is a single layer of flattened cells

← Diffusion of gases between an alveolus and a blood capillary in the lung

examiner tip

Partially permeable is preferable in exam answers to *semi-permeable*, but they both mean the same thing. Some substances can pass through a membrane and some (usually the larger molecules) can't.

Check your understanding

Tested

3 Are the following statements about diffusion true or false? *(4 marks)*
   a) It takes place down a diffusion gradient (from high concentration to low concentration).
   b) It occurs only across cell membranes.
   c) It is the process by which oxygen gets into cells.
   d) It is faster at low temperatures than at high temperatures.

4 a) What does partially permeable mean? *(1 mark)*
   b) What part of a cell is partially permeable? *(1 mark)*

5 List two factors that will affect the speed of diffusion of oxygen across a membrane. *(2 marks)*

Answers online   Test yourself online   Online

Animal organs

- All animals, by definition, are multicellular. There are no single-celled organisms that are classed as animals.
- Being multicellular allows for a **division of labour**: different cells can do different jobs.
- Animals develop from a single fertilised egg in which the cells divide to form an embryo. In the early embryo all the cells are the same, but as the embryo grows the cells **differentiate**, or specialise, so that they can perform particular roles.

Specialised cells for different jobs
Revised

All animal cells have particular jobs to do. Cells are adapted for a particular function —for example, nerve cells transmit impulses, muscle cells contract and cells in the lining of the gut make and secrete digestive enzymes.

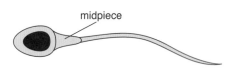
midpiece

Sperm — head contains DNA; midpiece contains mitochondria, which provide the energy for swimming; tail allows movement

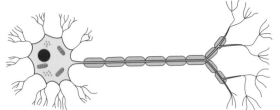
Nerve cells — are long and can transmit nerve impulses from one part of the body to another

Red blood cells — contain haemoglobin, which carries oxygen

↑ **Examples of specialised animal cells**

Specialised cells form tissues
Revised

A tissue is defined as a **collection of cells with similar structure and function**. Examples of tissues include:

- **muscle tissue**, which has one simple but vital function — it **contracts**
- **glandular tissue**, which **makes** and **secretes** substances such as mucus, enzymes and hormones
- **epithelial tissue**, which forms thin, flat sheets that act as **coverings** and **linings** for many parts of the body

    → Epithelial tissue covers or lines some parts of the body. The thin epithelium lining the lungs allows gases to diffuse through

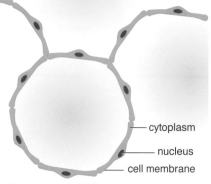
cytoplasm
nucleus
cell membrane

These epithelial cells around an alveolus are thin to allow gaseous exchange

Tissues work together to form organs
Revised

- An **organ** is a structure in the body that has a **specific function**.
- For example, the **heart** pumps blood, the **lungs** exchange gas.

Organs are made of tissues. For example, the **stomach** has:

- three layers of **muscle** tissue, to **churn** the contents
- **glandular** tissue to produce mucus and digestive **enzymes**
- **epithelial** tissue to **cover** the outside and the inside of the stomach

→ **Digestive system**

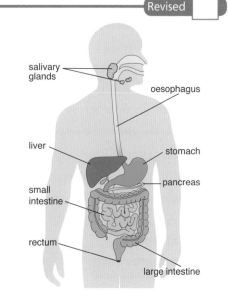
salivary glands
oesophagus
liver
stomach
pancreas
small intestine
rectum
large intestine

Organs work together to form systems

Systems are **collections** of organs that **work together** to perform a major function the body. The major systems of the human body include the digestive, circulatory, respiratory and nervous systems. For example, the digestive system includes:

- **glands**, such as the stomach, pancreas and salivary glands, which produce various digestive juices
- the **stomach** and **small intestine**, where **digestion** occurs
- the **liver**, which produces **bile**, a greenish digestive fluid important in the digestion of fats
- the small intestine, where the **absorption** of digested food occurs
- the **large intestine**, where water is absorbed from the undigested food, producing faeces

Plant organs

Plants are similar to animals in that they are all **multicellular**, and also have **specialised cells** that come together to form **tissues**. They have organs that include **roots**, **stems**, **leaves** and **flowers**.

Examples of plant tissues include:

- **epidermal** tissues, which cover most of the plant
- **mesophyll** tissue, which is found in the centre of leaves. The most important type of mesophyll cells are **palisade cells**, which carry out most of the plant's photosynthesis.
- **Xylem** and **phloem**, which are both composed of long, tubular cells used to transport substances around the plant. **Xylem** vessels take water and dissolved minerals from the roots to the upper parts of the plant, such as the leaves. **Phloem** vessels take the products of photosynthesis — mainly sugars — around the plant from where they are made (the leaves) to where they are needed (everywhere else).

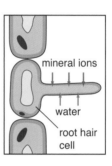 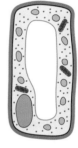

Root hair cell — large surface area to absorb water and minerals from the soil

Palisade cell (from a leaf) — contains lots of chloroplasts for maximum photosynthesis

↑ **Examples of specialised plant cells**

Check your understanding

6 List four organs that can be found in each of the following systems:

   a) digestive

   b) circulatory

   c) nervous

   d) female reproductive   *(4 marks: 1 for each correct set)*

7 Put the following in order of size, smallest first.   *(1 mark)*

   cell   tissue   organism   mitochondrion

   organ   population   system

Photosynthesis

It is easy to define an animal: they cannot make their own food, so they go in search of it. When they find it, they eat it, and digest it in a gut. This is true of all animals, from jellyfish to humans. In contrast, plants make their own food by **photosynthesis**, so they do not have to move.

Food from thin air
Revised

Photo means 'light' and synthesis means 'making'. The process of photosynthesis makes glucose using light energy, carbon dioxide gas from the air, and water.

Photosynthesis is the process that feeds the world, and pumps oxygen into the atmosphere at the same time. Photosynthesis can be summed up in the following equation:

Raw materials	Conditions	Products
carbon dioxide + water $6CO_2 + 6H_2O$	light energy chlorophyll enzymes $\rightarrow$	glucose + oxygen $C_6H_{12}O_6 + 6O_2$

→ **A section through a leaf — the plant's organ of photosynthesis. Most photosynthesis happens in the palisade layer, where there are lots of chloroplasts. When the plant is photosynthesising, carbon dioxide passes in and oxygen passes out through the stomata**

- **Light energy** from the Sun is trapped by molecules of **chlorophyll**, and the energy is used to turn **carbon dioxide** into **glucose**.
- **Oxygen** is released as a by-product.

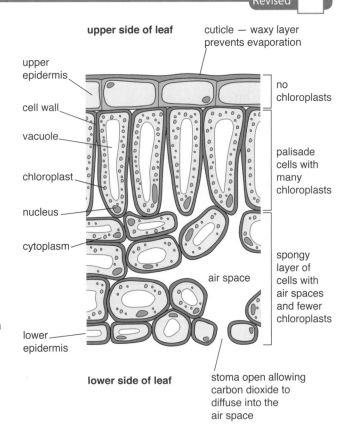

upper side of leaf

cuticle — waxy layer prevents evaporation

upper epidermis

cell wall

vacuole

chloroplast

nucleus

cytoplasm

lower epidermis

lower side of leaf

no chloroplasts

palisade cells with many chloroplasts

air space

spongy layer of cells with air spaces and fewer chloroplasts

stoma open allowing carbon dioxide to diffuse into the air space

What limits the rate of photosynthesis?
Revised

Plants have a few basic needs: light, carbon dioxide, water, minerals and warmth.

The one that is in shortest supply is called the **limiting factor**. Give the plant more of its limiting factor and it will grow faster, until something else becomes limiting. For example, if light is limiting, increasing the light intensity will increase the rate of photosynthesis until something else becomes limiting, such as the supply of carbon dioxide.

- If there is not enough **light**, there is not enough **energy** to **activate** the **chlorophyll**.
- If there is not enough **carbon dioxide**, the plant is short of the raw material it needs to make **glucose**.
- **Temperature** is often a limiting factor — in cold regions plants grow more slowly.
- Photosynthesis reactions are controlled by **enzymes**, which have an optimum temperature. The lower the temperature, the slower the rate of photosynthesis and the slower the growth.
- Above a certain temperature, the enzymes are **denatured** and do not work, so the rate of photosynthesis decreases.

examiner tip
Limiting factors are a favourite exam topic.

What do plants use glucose for?

- Plants use glucose for **respiration** to release energy.
- Plants also use glucose to make various other compounds, such as **starch**, cellulose, lipids and protein.

Starch is produced in the leaves, so a **starch test** using **iodine solution** shows if a leaf has been photosynthesising. Plants store glucose as starch because it is insoluble, and does not make the cytoplasm too concentrated. Starch is converted into sucrose before being transported to other parts of the plant, such as roots, flowers or fruits.

For a plant to make **proteins** and other vital substances such as **DNA**, it needs a supply of mineral ions from the soil. **Nitrate** is needed to make proteins, while DNA requires **nitrate** and **phosphate**.

Check your understanding

8 State two reasons why photosynthesis is vital to life on Earth.

(2 marks)

9 Write out the overall equation for photosynthesis, in words.

(3 marks)

10 If a cell's cytoplasm contains a lot of glucose, it will absorb a lot of water due to osmosis. Why does the cell not have the same problem with starch?

(1 mark)

11 Match up each limiting factor to what that factor is needed for.

(6 marks)

Limiting factor	Needed for
a) Carbon dioxide	**(i)** Making proteins
b) Magnesium	**(ii)** The energy required for photosynthesis
c) Nitrate	**(iii)** Making sugars via photosynthesis
d) Light	**(iv)** Trapping light energy
e) Warmth	**(v)** Making chlorophyll
f) Chlorophyll	**(vi)** Enzyme-controlled reactions

12 A student was asked to investigate the effect of temperature on the rate of photosynthesis in a piece of pondweed.

a) Suggest how the student could measure the rate of photosynthesis. *(1 mark)*

b) What is the dependent variable in this investigation? *(1 mark)*

c) What is the independent variable? *(1 mark)*

d) List three variables that the student would have to control to make it a fair test. *(3 marks)*

e) Sketch a graph to show the effect of temperature on rate of photosynthesis. *(3 marks)*

13 Give two reasons why the growing season in the Arctic circle is so short. *(2 marks)*

Answers online　　Test yourself online

Distribution of organisms

There are many factors that affect the distribution of organisms. Generally, they can be divided into **living** and **non living**.

Living, or **biotic**, factors include:

● availability of **food**

● **predators**

● **competition** for territory, mates, nesting sites etc.

● **disease**

Non-living, or **abiotic**, factors include:

● **temperature** — all living things rely on **enzymes**, which are very temperature sensitive

● **wind/air movements** —these affect the rate of evaporation/water loss, especially in plants

● **light** — essential for photosynthesis

● **water** — essential for all organisms

● **oxygen** — this is not normally a problem in air, but water holds much less oxygen and so aquatic organisms can struggle to get enough

● **carbon dioxide** — essential for photosynthesis

● **nutrients** — plants need a supply of **ions** such as **nitrate** from the surrounding water or soil

Measuring distribution of plants

Revised

A lot of fieldwork is done on plants and other organisms that cannot move, or that move slowly, such as barnacles and limpets on a rocky shore.

It is rarely possible to count all the organisms in a particular area — it would take far too long and you would lose the will to live. To overcome the problem, we have **sampling**.

There are two key sampling techniques: **quadrats** and **transects**.

● Quadrats are used to **compare two different areas**, such as two sides of a tree, or a mown and un-mown field.

● Transects are lines that are used to sample **areas of gradual change**, such as the changes in seaweeds and animals up a rocky shoreline, or how sand dunes change as you go inland from the sea.

● Quadrats can be thought of as random patches of land that represent the whole area, while transects are systematic samples taken along a line.

→ **Quadrats can be used to compare the lichen and moss distribution on different trees, or on the north and south facing sides of trees**

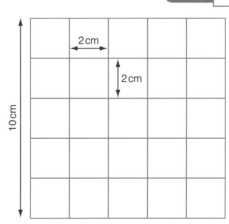
a 10cm quadrat marked on a clear sheet

1. The quadrat is placed in a few randomly chosen places.
2. The number of lichens within the quadrat, or the area covered, is counted.
3. Only about 2% of the total site needs to be sampled to give an average.

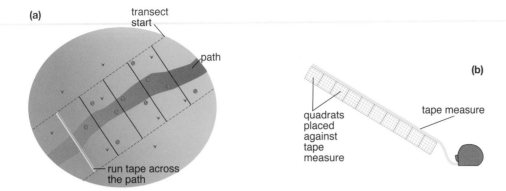

(a) transect start, path, run tape across the path

(b) tape measure, quadrats placed against tape measure

↑ Some plants can survive being trampled, and some cannot. To investigate the effect of trampling on the diversity of species, you could run a transect across a path a). By placing a quadrat continuously along the transect b), you can count the number of different species present

Field studies can give **qualitative** or **quantitative** data:

● Qualitative data will tell you **what** species are present.

● Quantitative data involve numbers, so might tell you **how many** of each species are present, or the percentage cover of a particular species.

Monitoring environmental change

Revised

Once you have established that there is a difference in the distribution of organisms, the next step is to try to find out why.

Differences may be due to biotic or abiotic factors. If they are due to abiotic factors, sensors can be used to used to monitor the levels in different places. For example, if you throw quadrats on either side of a hedge, you might find that there is a greater species diversity on the north-facing side. Sensors could be used to measure levels of light, temperature, soil nutrients, soil pH, wind and humidity.

Many sensors are in place around the world to monitor climate change. Sea temperatures, pH levels, wind speed, rainfall and many other measurements are taken to find out the extent of environmental change in different parts of the world. Many of these sensors are remote, i.e. unmanned, and automatically send data to labs around the world for analysis.

Check your understanding

Tested

14 Lichens are organisms that are very sensitive to air pollution. A fieldwork survey used quadrats to estimate the percentage cover of lichens growing on tree trunks in different areas of the school grounds.

   a) Suggest two different areas to investigate, giving a hypothesis for why there might be a variation between the two areas.   *(2 marks)*

   b) Describe a method to determine where a sample should be taken.   *(1 mark)*

   c) Explain why further samples should be taken.   *(2 marks)*

Answers online ── **Test yourself online**

Online

Proteins

Proteins are large organic molecules that play a central role in living organisms. In the human body, proteins include:

- the fibres that make your **muscles** contract
- **collagen**, which gives strength to tendons, bones and other tough tissues
- **keratin**, which gives strength to skin, hair, nails and hooves
- Some **hormones**, such as insulin
- **antibodies** — a vital part of the immune system
- **enzymes** — see the next section

Proteins are made from amino acids `Revised ☐`

Proteins are very large molecules, and are made up from smaller units called **amino acids**.

- There are 20 different amino acids, which can be joined together in any order.
- Different proteins are made from different **sequences** of amino acids.
- Amino acids form long chains that fold and bend to produce differently shaped protein molecules.
- This shape is vital to the functioning of the protein.

For example, an antibody molecule has to be exactly the right shape to combine with a particular antigen. An enzyme has to be exactly the right shape to catalyse a particular reaction.

There is an infinite number of different proteins. This is vital to life because there can be one enzyme for every reaction, one antibody for each disease.

Enzymes

Enzymes are:

- substances that speed up reactions in living things — they are biological **catalysts**
- very **specific** — there is a different enzyme for every reaction
- **delicate** — they will not work if the conditions are not right
- large, complex **protein** molecules

You can often tell what an enzyme does from its name: protease enzymes digest protein, maltase digests maltose, and so on.

You need to be able to describe the use of enzymes in three different areas:

1 enzymes that work **inside** cells, controlling thousands of reactions

2 enzymes that work **outside** cells, in our **gut** where they **digest** our food (see page 51)

3 enzymes that work for us in the home or industry (see pages 53–54)

> **examiner tip**
>
> Enzymes are not living things; they are molecules. They cannot be killed.
>
> Not all enzymes are digestive enzymes; most are not.
>
> The substrate and the active site are **not the same shape**; they are **complementary**.

Enzyme shape is vital for its function

Revised

- The **substrate** is what the enzyme works on.
- The **active site** is a region on the surface of the enzyme that matches the substrate.

Enzymes, like all proteins, are made up of long chains of **amino acids**. Each chain always **folds** in a particular way to form a very precise shape. If anything changes this shape, such as high temperature, the enzyme will not work — the enzyme has been **denatured**.

Enzymes also have an **optimum pH**, where they work best. For enzymes that work inside cells this pH is around neutral. Some digestive enzymes have extremes of optimum pH; for example, protein-digesting enzymes in the stomach have an optimum pH of about 2–3 (strong acid).

chain of amino acids   overall shape of molecule

↑ The enzyme molecule is a long chain of amino acids, bent and folded into an exact shape. The active site will fit only one particular substrate

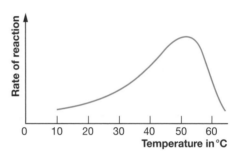

↑ All enzymes have an optimum temperature, where the rate of reaction is fastest. For this enzyme the optimum is about 50°C

At correct temperature

active site — this region matches the substrate

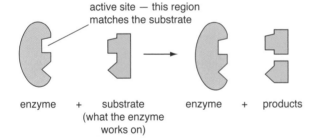

enzyme   +   substrate   enzyme   +   products
                (what the enzyme
                works on)

Temperature too high

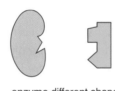

enzyme different shape,
i.e. denatured

↑ High temperatures make enzyme molecules vibrate so much that their shape changes, stopping them from working. When the shape changes, the substrate will not fit into the active site: the enzyme has been denatured

Some enzymes work inside cells

Revised

There are thousands of different enzymes, and most of them work inside cells. Examples include enzymes that catalyse processes such as **respiration**, **protein synthesis** and **photosynthesis**.

In **protein synthesis**, enzymes inside cells catalyse the reactions that build up amino acids into proteins.

examiner tip

Generally, enzymes are controlled by hormones. Think of enzymes as workers and hormones as bosses, telling enzymes when to work and when to stop.

Speed of enzyme activity

Revised

There are several different factors that affect the speed of an enzyme-controlled reaction:

- The **temperature** — the higher the temperature, the faster the enzyme and substrate molecules bounce around and collide, so the higher the rate of reaction. Above a certain temperature, however, the shape of the enzyme molecule changes — it becomes **denatured** and its activity is permanently stopped.

- The **pH**. All enzymes have an optimum pH at which they work best.

- The **amount of substrate**.

- The **amount of enzyme**.

- Presence of **inhibitors** — substances that slow down or stop enzyme activity.

Check your understanding

Tested

15 Explain how we can eat a piece of meat such as chicken, and use it to make completely different proteins in our bodies. *(3 marks)*

16 Use the words below to complete the text that follows.

denatured   protein   active site   specific   substrate

Enzymes are _____ molecules that have a precise shape. There is a region on the surface of the enzyme called the _____ _____ , into which the _____ fits. Enzymes are described as _____ because they only catalyse one particular reaction. If the temperature becomes too high, the enzyme molecule will vibrate, lose its shape and will be _____ . *(5 marks)*

17 An enzyme works on a particular substrate, turning it into product. The graph shows how one particular enzyme digested 1 g of substrate at 20°C. The optimum temperature for this enzyme is 40°C.

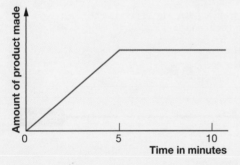

a) Explain what happens, in terms of substrate and product:

  (i) from 0 to 5 minutes

  (ii) from 5 to 10 minutes *(2 marks)*

b) Suggest how the graph would change if we repeated the test:

  (i) at 20°C higher

  (ii) with 2 g of substrate

  (iii) at 40°C higher *(3 marks)*

Answers online   Test yourself online

Online

Enzymes and digestion

Digestive enzymes work outside cells

Revised

The digestive system is basically a long tube that runs from mouth to anus. It can be thought of as a food-processing conveyor belt. Different stages of digestion occur in different parts.

Food consists of large molecules such as proteins, carbohydrates and lipids, along with simpler ones like vitamins, salts and water.

● The large molecules are too big to pass through the intestine wall into the blood.

● **Specialised cells** in **glands** in the lining of the gut secrete **digestive enzymes**.

● These enzymes pass into the gut where they **break the food down** into **smaller**, **soluble** molecules.

● As food passes along the gut the digested molecules are **absorbed** into the blood. What is left is all the stuff that cannot be absorbed: plant fibre, bacteria, dead cells, some digestive juices and water. That is what comes out at the other end.

examiner tip

Food that we have eaten is not actually part of our bodies until it has been absorbed. The content of the gut is a nasty mixture of half-digested food and bacteria that would cause all sorts of problems if it were actually circulating in our blood, inside our bodies.

Learn where each enzyme is produced and where it acts. Remember that other substances are needed to optimise the pH for each enzyme, so learn where these other substances are produced and released when they are needed.

Salivary glands — saliva contains **amylase**, an enzyme that breaks down starch into sugars. This is the first stage in the chemical digestion of food.

Mouth — chewing is the mechanical breakdown of food, making it smaller and mixing it with **saliva** so it can be swallowed.

Liver — this secretes **bile**. Bile is stored in the gall bladder then released in the small intestine. It contains no enzymes but it does **emulsify** lipids — it turns large droplets into smaller ones, giving a large surface area for the lipase enzymes to work on.

Stomach — the thick muscular wall churns food and secretes **gastric juice**. This contains **hydrochloric acid** and a **protease** enzyme. The enzyme begins the digestion of protein, breaking it down into short chains of amino acids. The stomach has a much lower pH than the rest of the gut.

← Digestive system

Pancreas — pancreatic juice contains lots of different enzymes including **protease, lipase** and **amylase**.

Colon — or large intestine. Region of water absorption. Faeces are stored in the rectum before being removed through the anus.

Small intestine — the main region of **digestion** and **absorption**. The gut wall secretes protease, lipase and amylase. Digestive juices from the pancreas and liver are also added here.

Providing the optimum pH

Different enzymes work best at different pH values:

- Protease enzymes released in the stomach work most effectively in very **acidic** conditions, so glands in the stomach wall secrete **hydrochloric acid** at the same time as the protease enzymes.

- The protease, amylase and lipase enzymes released in the small intestine work most effectively in **mildly alkaline** conditions. Bile **neutralises** the acid that was added to food in the stomach.

→ Overall, our digestive enzymes must turn the big three food types (carbohydrates, lipids and proteins) into small, simple molecules that we can absorb.

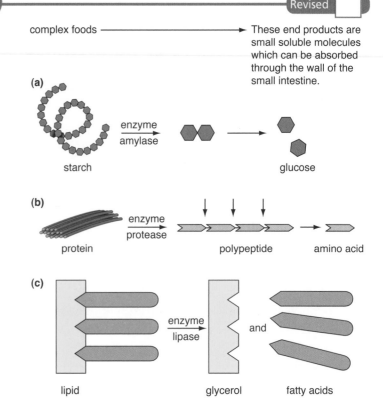

complex foods → These end products are small soluble molecules which can be absorbed through the wall of the small intestine.

(a) starch — enzyme amylase → glucose

(b) protein — enzyme protease → polypeptide → amino acid

(c) lipid — enzyme lipase → glycerol and fatty acids

Breaking down large molecules

Digestive enzymes break down large molecules into smaller molecules.

- **Amylase** is the enzyme that breaks down starch into glucose molecules.

- **Protease enzymes** break down proteins into amino acids. This is a two-stage process. Proteins are first digested into short chains of amino acids, then into individual amino acids.

- Lipids (fats/oils) are broken down by **lipase** enzymes into fatty acids and glycerol.

Check your understanding

18 a) Name three parts of the digestive system that produce amylase. *(3 marks)*

   b) Name two parts of the digestive system that produce lipase. *(2 marks)*

   c) Explain how proteins are digested. *(2 marks)*

   d) What happens to 'food' molecules once they have been digested? *(1 mark)*

19 Some people with liver damage cannot produce bile. How would their digestion be affected?

   (2 marks)

20 Match up the region of the gut with the most appropriate description. *(6 marks)*

Region	Description
a) Oesophagus	**(i)** Muscular bag that stores and churns food
b) Stomach	**(ii)** Main region of digestion and absorption
c) Pancreas	**(iii)** Muscular tube that allows swallowing
d) Gall bladder	**(iv)** Absorbs water from food
e) Small intestine	**(v)** Stores bile
f) Large intestine	**(vi)** Produces digestive enzymes and hormones

Uses of enzymes

Enzymes have many uses in the home, industry and medicine because they can make **particular reactions** happen, usually very **quickly** and **without the enzyme being used up** in the process.

We have already seen (in decay processes, page 28) that some **microorganisms** make and secrete enzymes. This is useful because we can collect these enzymes without having to break open the cells of the microorganism. Purifying the enzyme — separating it from the microbe that made it — is easy and cheap.

Laundry detergents   Revised ☐

Biological washing powders and dishwasher detergents have enzymes added in order to break down fat, protein and starch stains.

Biological detergents are more effective at low temperatures, so less energy is used in heating water.

Enzymes in the food industry   Revised ☐

There are many uses for enzymes in the food industry. For example, **fructose** is a sweeter sugar than glucose or sucrose. **High-fructose syrup** is in big demand by makers of sweet foods and drinks because they can use less, so it is cheaper. It is also easier to make **low-calorie** foods using fructose.

Starting with starch, which is a cheap and abundant substance, fructose is made in a two-stage process:

1 The carbohydrases **amylase** and **maltase** are used to convert **starch** into **glucose** syrup.

2 **Isomerase** is used to convert **glucose** syrup into **fructose** (glucose and fructose are isomers of each other — they have the same atoms but a slightly different structure).

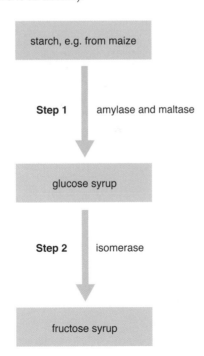

← Steps in using enzymes to make fructose syrup

examiner tip

Exam questions often contain enzymes you will never have heard of. Don't panic: the questions are just testing your knowledge of enzymes in general.

Proteases are used to 'pre-digest' the protein in some baby foods, making them softer and easier for the baby's gut to absorb.

The table shows some examples of useful enzymes.

Enzyme type	What it does	Why it is useful
Protease	Digests protein stains such as gravy, egg and blood	Washing powders
	Digests protein in baby food	Makes baby food easier to chew and digest
Lipase	Digests lipid stains such as cooking oil	Washing powders
Carbohydrase	Digests starch into glucose	Step 1 in making high-fructose syrup
Isomerase	Converts glucose into fructose	Step 2 in making high-fructose syrup

Why enzymes are used in industry
Revised

- Enzymes are used to make reactions happen at normal temperatures and pressures; otherwise, expensive equipment that uses large amounts of energy would be needed.

- The enzyme is not used up, so it can be used again and again.

However, each enzyme works on only one substrate, and in particular pH and temperature conditions. Finding new enzymes to work on a particular substrate in an industrial process can be very costly.

Check your understanding
Tested

21 Some bacteria have been found living in hot volcanic springs, where the water is almost boiling. The enzymes in these bacteria are thermostable, i.e. able to work at temperatures that would denature most enzymes. Suggest why enzymes from these bacteria are particularly useful for washing powders. *(2 marks)*

22 Explain one possible advantage and one disadvantage of using washing powders that contains enzymes. *(2 marks)*

23 Why is it easier to make low-calorie food with fructose than with glucose? *(2 marks)*

Answers online — Test yourself online
Online

Aerobic respiration

- Respiration is one of the seven signs of life. It is a process that goes on in **all cells**, in **all organisms**, **all of the time**.
- Animals, plants, fungi and bacteria respire all the time, **all day and all night**.
- The only cells that do not respire are either dead or dormant (such as seeds in winter).

The vital point about respiration is that it **releases energy from organic molecules** such as glucose and lipid (fat). In fact, many organic molecules can be respired to release energy, but humans normally use glucose, and resort to respiring fat when the glucose runs out.

Respiration and energy release
Revised

The energy released in respiration can be used for many different things, including:

- the **contraction of muscles**
- keeping a **stable body temperature** in mammals and birds
- **growth** — building up small molecules into larger ones, for instance building **amino acids** up into **proteins**
- **movement** of substances in or out of cells by **active transport** (page 65)

There are two basic ways to respire. **Aerobic** respiration uses oxygen, while **anaerobic** respiration does not need oxygen.

Aerobic respiration can be summarised as follows:

glucose + oxygen → carbon dioxide + water + energy

Most of our cells respire aerobically most of the time. It is an efficient process that gets a lot of the energy out of glucose. It can be thought of as the **complete** breakdown of glucose, to get **all** the energy out.

Most of the reactions of aerobic respiration take place inside the **mitochondria**.

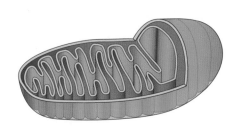

↑ A mitochondrion

Respiration and exercise
Revised

The energy to move our muscles comes from respiration. When muscles are working hard:

- a lot of **oxygen** and **glucose** gets used up
- a lot of carbon dioxide and heat gets produced

So, during exercise, the brain can detect increased **carbon dioxide** levels and responds by:

- increasing the **heart rate**
- increasing the **ventilation rate** — we breathe **deeper** and **more frequently**

The increased blood flow to the muscles delivers more oxygen and glucose (or 'blood sugar') and takes away the excess carbon dioxide and heat.

When we are short of glucose we can instantly get more by breaking down a storage substance called **glycogen**.

● Glycogen is stored in the **liver** and **muscles**.

● It is simply a large molecule made from thousands of glucose molecules bonded together.

● When required, **enzymes** break down the glycogen and release more glucose for the muscles to keep working.

● We build up our glycogen stores again when we eat carbohydrate foods such as starch.

Anaerobic respiration

● Anaerobic respiration is **respiration without oxygen**.

● It is shown as a simple equation:

glucose → lactate + energy

● It is simpler and quicker then aerobic respiration, but it does not release as much energy.

● It can be thought of as the **incomplete breakdown** of glucose. Some of the energy in glucose is released, but to get it all out requires a supply of oxygen.

● The incomplete breakdown of glucose produces **lactic acid**.

Muscles can respire anaerobically when they run short of oxygen. This allows them to continue to contract even when the oxygen supply runs out.

Our muscles can only work for a short time without oxygen, because there is a painful build up of lactic acid (lactate), which changes the pH in muscles.

Once exercise has stopped, blood will deliver oxygen to the muscles, which is used to break down the lactate. The amount of oxygen needed to break down the lactate into carbon dioxide and water is known as the **oxygen debt**.

Yeast can respire either aerobically or anaerobically. If yeast cells are deprived of oxygen they respire anaerobically but will not produce lactate. Instead, they produce **ethanol** (alcohol) and **carbon dioxide**.

> **examiner tip**
> Lactic acid and lactate are the same thing.

Check your understanding — Tested

24 How would you measure the rate of respiration in maggots? *(1 mark)*

25 Suggest why a marathon runner might want to eat a lot of pasta-based food before a race. *(2 marks)*

26 The longest race that is classed as a sprint is 400 m. Suggest why. *(2 marks)*

Answers online — **Test yourself online** — Online

Cell division

What are chromosomes and genes?

Revised

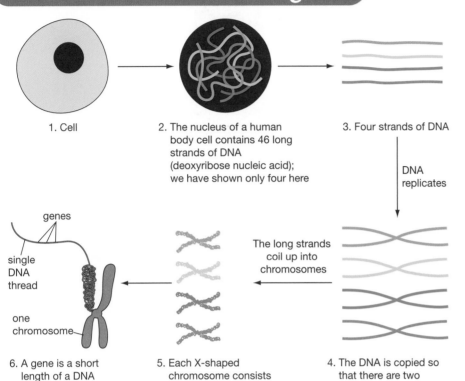

1. Cell

2. The nucleus of a human body cell contains 46 long strands of DNA (deoxyribose nucleic acid); we have shown only four here

3. Four strands of DNA

DNA replicates

The long strands coil up into chromosomes

genes

single DNA thread

one chromosome

6. A gene is a short length of a DNA strand

5. Each X-shaped chromosome consists of two identical strands

4. The DNA is copied so that there are two identical strands, joined together; there are now four double strands

> **examiner tip**
>
> Many students state that all organisms have 46 chromosomes. Humans do, but most species have a different number. Chickens, for example, have 78, while cats have 38.

Chromosomes and cell division

Revised

To understand why everyone has genetic differences (except identical twins), we need to take a close look at cells and **cell division**.

Your body is made of two types of cell:

● **Body cells** — each one contains two sets of chromosomes, 46 in all. The vast majority of our cells are body cells.

● **Sex cells**, or **gametes** — these are **spermatozoa** (sperm) in males, and **ova** (eggs) in females. These cells have just one set of chromosomes, 23 in all.

To make the two types of cell, we need two different types of cell division:

● **Mitosis** is straightforward cell division in which one body cell copies itself into two identical body cells. This is how new cells are made when we grow and develop, and replace old cells. (It is also how organisms are cloned — see pages 32–33)

● **Meiosis** is a special, more complicated type of cell division that makes eggs (in the **ovaries**) and sperm (in the **testes**).

> **examiner tip**
>
> Asexual reproduction just involves mitosis. There is no mixing of genes and all new individuals are identical, i.e. clones.

Meiosis does two vital things:

1 It **halves the number of chromosomes to 23**, so that the full number of 46 chromosomes will be restored when a sperm **fertilises** the egg.

2 It **shuffles the genes** so that every new cell is different.

In meiosis, the cell divides **twice**, so **four** gametes are produced, each with a single set of chromosomes. This is why body cells have two sets of genetic information but gametes have only one set.

Where do babies come from?

Revised

- When a sperm fertilises an egg, the result is a single diploid cell called a **zygote**.
- This cell divides repeatedly by **mitosis** to produce a ball of cells, the early **embryo**.
- At this stage the cells are all the same — they are **undifferentiated**.
- But they are **stem cells** with the **potential** to differentiate into **any** specialised cell.
- All the cells in the embryo contain the **same genes**, but the cells specialise, or **differentiate**, when **different genes** are activated.

The stem cells in the early embryo are of great interest to scientists because it may be possible to control their differentiation. If this can be done there is the potential to treat many different conditions, for example:

- repair of spinal cord injuries in cases of **paralysis**. Nerve cells do not normally repair themselves very well.
- making replacement insulin-making cells for **diabetics**
- making replacement brain cells for sufferers of **Parkinson's** or **Alzheimer's** disease.

As the animal embryo develops, most of its cells lose the power to differentiate. Many cells can divide, but they can only make new cells of the same type. Apart from those in the embryo, there are two types of stem cell currently of interest to scientists:

- **umbilical cord** — the connection between the baby and the placenta
- **bone marrow**

These stem cells are not quite as versatile as embryonic stem cells — they cannot turn into as many different cell types. However, they might still prove very useful in medical treatments.

Plant cells, in contrast, remain stem cells for **all their life**. A cell taken from anywhere on a plant — a leaf for example — can be made to grow into a completely new plant, given the right conditions.

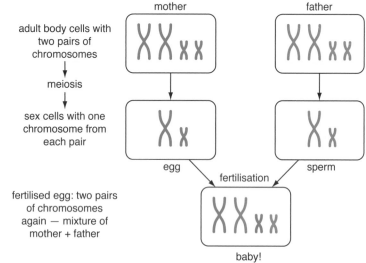

↑ Meiosis takes one from each pair of chromosomes at random, and then shuffles the genes so that no two daughter cells are identical. Eggs and sperm contain a single set of chromosomes. At fertilisation, egg meets sperm and the two sets of chromosomes match up, producing a new individual with 46 chromosomes. The fertilised egg is a zygote

examiner tip

Remember MEIOSIS 'Makes Eggs In Ovaries and Sperm In the Scrotum'. You will never confuse it with mitosis again.

Check your understanding

Tested

27 Put the following in order of size, smallest first. *(1 mark)*

chromosome   nucleus   cell   gene   tissue

28 a) When a cell undergoes meiosis, what is the result? *(3 marks)*

   b) If the original cell has 34 chromosomes, how many will the daughter cells have after meiosis? *(1 mark)*

29 Cells in the skin can divide by mitosis again and again. Why are they not classed as stem cells? *(1 mark)*

30 Why might people object to the use of stem cells? *(2 marks)*

Answers online   **Test yourself online**   Online

Genetic variation

Why reproduce sexually?

Revised

- Sexual reproduction has evolved because it has great **survival value**.
- Meiosis **shuffles** the genes so that **every sperm** and **every egg** is **different**.
- Any sperm can fertilise any egg, so that each new individual created is genetically different from any that have gone before.
- If all individuals are different, there is much more chance that some individuals will have what it takes to survive.
- The only way to get individuals that are **genetically identical** to the parent is by **asexual reproduction**, where the offspring are produced by just **one** parent. These individuals are all **clones**.

Boy or girl?

Revised

It is all about chromosomes:

- The sex of a baby is decided at the moment of fertilisation, and depends on which sperm gets there first.
- The body cells of males have a pair of sex chromosomes, one X and one Y.
- Meiosis puts one chromosome from each pair into each sperm.
- So half the sperm will get an X chromosome, and half a Y.
- All body cells in girls have two X chromosomes, so all of their eggs contain one X.
- **If a Y-carrying sperm fertilises the egg, you get a boy.**
- **If it's an X-carrying sperm, you get a baby girl.**
- There is a 50:50 chance.

> **examiner tip**
>
> In questions about sex determination, many candidates write about 'male sperm' and 'female sperm'. They should be referred to as 'Y-carrying' or 'X-carrying' sperm.

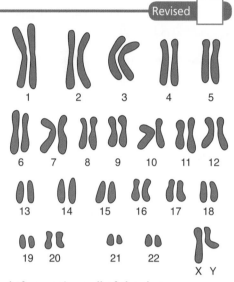

↑ If you pair up all of the chromosomes in a human body cell and then arrange them in order of size, this is what you get. The last pair are the sex chromosomes. The large one is an X chromosome and the smaller one is a Y chromosome. This individual has one of each, so we know he is male. Females have two X chromosomes

Genes or alleles?

Revised

To understand how features are passed on from one generation to the next, we can look at those that are controlled by **single genes**.

First, some vital definitions:

- **Gene**: a short piece of DNA that **codes** for making a particular **protein**.

  The information in the gene tells the cell to put amino acids together in a particular order, so a particular protein is made.

- **Allele**: an alternative form of a gene. In peas the flower colour gene has two alleles, red and white. We need to give these alleles symbols, so let us call the red allele **R** and the white allele **r** (it is better to stick to different forms of the **same letter** to show different forms of the **same gene**).

- A **dominant** allele is one that, if present, will be shown in the phenotype (the observable features of an organism — see below).

- A **recessive** allele is shown only if the dominant allele is absent.

In reality, it is not the allele that makes the flower red. The allele will make an enzyme that makes the red colouring.

Genetic diagrams

Pea plants, like most organisms, have two versions of each gene (one on each chromosome). So they will have two alleles that code for flower colour. This gives us three possibilities:

● Both alleles code for red flowers, so we say the **genotype** is **RR**.
● One codes for red and one for white: genotype **Rr**.
● Both alleles code for white: genotype **rr**.

Meiosis will **separate** theses alleles so that only **one copy** goes into the gamete.

In an organism with a genotype of **Rr**, half of the gametes will contain the **R** allele and the other half will contain the **r** allele.

If the genotype is **RR** or **rr**, all the gametes will contain the same allele.

The observable features of an organism are its **phenotype**. We can predict that **RR** plants will have red flowers, and that **rr** plants will have white flowers, but what about Rr? Whether they are red or white depends on which characteristic is **dominant**.

In this case red is dominant, so **Rr** plants will be **red**. This means that the **r** allele will be hidden, or **carried** (Example A below).

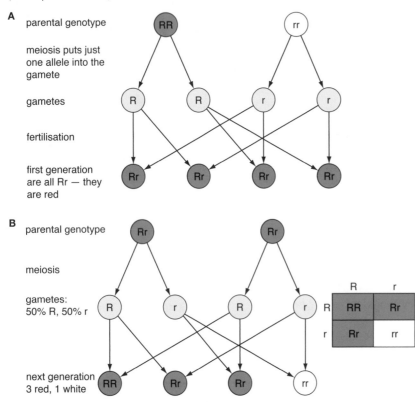

← Example A: a cross between a red plant and a white plant

What happens if we then breed the first generation together? This time both parent plants are red, but they are carrying the recessive r allele (example B)

Check your understanding

31 Explain why roughly equal numbers of male and female babies are born. *(5 marks)*

32 A long-coated ferret is mated with a short-coated ferret. All of the babies are long coated.

a) Define the words phenotype and genotype. *(2 marks)*

b) Is the long coat or short coat feature dominant? *(1 mark)*

c) If the letter L and l represent the two coat length alleles, state the genotype of the short-coated parent and the possible genotype of the long-coated parent. *(3 marks)*

d) Construct a genetic diagram to predict the outcome of a mating between two of these baby ferrets. *(3 marks)*

e) Suppose you have a long-coated ferret and you want to know if it is carrying a gene for short coat. Suggest a breeding experiment you could carry out, and explain what the results would tell you.

(3 marks)

Answers online — Test yourself online

Genetic disorders

- Some diseases, like the flu and typhoid, are **communicable**. They can be caught because they are caused by **pathogens**.
- Some diseases, like heart disease and cancer, are caused by a complex mixture of **genes** and **lifestyle**.
- But some diseases are simply due to **genes**.

Cystic fibrosis
Revised

Some of the clearest examples of human genetic inheritance are seen in cases of **genetic disease**. **Cystic fibrosis** is one of the commonest genetic diseases. Symptoms include a build up of sticky mucus in the lungs. This makes breathing difficult, and bacterial infections are common.

- The underlying cause of cystic fibrosis is a **recessive allele**, which makes a **faulty protein**.
- In healthy people, the normal allele makes a protein that works in cell membranes, helping to pump salt and water.
- In people with cystic fibrosis, the protein does not work and no salt or water is pumped into the mucus, leaving it stickier and harder to move than normal mucus.

Looking at the possible genotypes:

- Most people have two healthy alleles; their genotype is **FF**.
- About 1 in 20 people carry a faulty **f** allele; their genotype is **Ff**. They are healthy because they have one dominant F allele, and that is all they need to make the protein. However, they are **carriers** of the faulty allele, and so could pass it on to their children.
- People with cystic fibrosis have two faulty alleles, **ff**.

Polydactyly
Revised

Polydactyly is the perfectly harmless condition of having an extra finger or toe. It is nothing more than a curiosity but it can be quite distressing to a child who has the condition. It is caused by a dominant allele, so anyone who does not have the condition can be sure that they do not carry the allele.

Individuals who do carry the allele have a 50% chance of passing it on to their children.

DNA testing
Revised

DNA testing or profiling — commonly called **DNA fingerprinting** in the media — was developed in the 1980s. It works on the principle that everyone (except identical twins) has **unique** DNA. It has many uses:

- At **crime scenes**, suspects can be identified from their blood, semen or skin, or even from mucus from a sneeze.

● **Paternity testing** — to identify a child's father. This can also work for other family links, for example to identify brothers, sisters or cousins.

● To find out whether an individual is carrying a particular gene or allele, for example for cystic fibrosis.

Will all future babies be born healthy?

Revised

If a couple suspect that their baby might carry a genetic disease, **embryo screening** (also called pre-implantation genetic diagnosis, PIGD) can help.

IVF (in vitro fertilisation, 'test-tube baby') technology is used, with the addition of genetic testing.

1 Eggs are collected from the woman.

2 Eggs are fertilised by the sperm.

3 Embryos start to develop in vitro (in glass, i.e. in the laboratory).

4 A cell is taken from the embryo and tested for the genetic disease. (You can take a few cells from embryos without doing them any lasting damage.)

5 Only healthy embryos are implanted into the mother.

examiner tip

You should be able to discuss the benefits of, and ethical concerns about, embryo screening, and summarise views for and against.

Embryo screening has benefits but there are ethical and moral objections. The use of the technology is a difficult personal decision.

For	Against
● Avoids the difficult choice of abortion if the embryo has inherited the genetic disease ● Not fair to bring a child into the world with a disorder if that could have been prevented ● Parents would suffer distress of bringing up a child with a serious illness ● Avoids the chance of a child developing a serious illness in middle age ● Saves cost of treatment later in life	● Embryos are destroyed — parents should just accept the baby, whatever its health ● Who has the right to decide which embryos live or die? ● Parents might want to screen embryos for non-medical reasons, e.g. to have a male child or even to have a child with a particular gene for eye or hair colour ● IVF with embryo screening is expensive

Check your understanding

Tested

33 A pregnant woman finds out that both she and her partner carry the cystic fibrosis allele.

a) Is this allele dominant or recessive? *(1 mark)*

b) What are the chances that the baby will be born with the disease? *(1 mark)*

c) What can she do to find out if the baby will have the disease? *(1 mark)*

d) What choices will she face if she finds out that the baby does have the disease? *(2 marks)*

Answers online ——— Test yourself online

Online

Old and new species

We are not sure how life began

Revised

- It is important to appreciate that the Earth is **extremely old**. Our current best estimate is that our planet is **4.6 billion years** (4 600 000 000 years) old.
- Scientists think life on Earth began about 3.9 billion years ago.
- There are many different theories about how life began. Some are based on religion, others are more firmly rooted in **evidence-based science**.
- The first living things were like bacteria — tiny and **soft-bodied** — so they left little **evidence** behind. As organisms evolved bigger and more complex bodies, more started to become **fossilised**.

Fossils help piece it all together

Revised

Fossils are evidence that a species once existed. It is often the hard parts of the organism that get fossilised, such as a skeleton or shell, a leaf or root. Sometimes it is just evidence of the organism's lifestyle, such as a nest, burrow, footprint or even a pile of dung.

Organisms are fossilised when **hard tissues** such as the skeleton **do not rot**, or if the conditions for decay are not right. This can happen if there is not enough oxygen for the decomposers, or the pH is wrong. The original tissues can become replaced by rock in a process called **mineralisation**.

Different layers of rock have different fossils, so we can often figure out what lived with what, and work out a **timeline**. We know that there were fish before amphibians, for instance.

Where did all the dinosaurs go?

Revised

Dinosaurs are just one small group of animals among the many that are now **extinct**. The species alive on Earth today are far outnumbered by those that have died out.

How does extinction happen? A species may become extinct if:

- a **predator** or **disease** kills them all
- a new species **outcompetes** them, usually for food
- the climate or environment **changes too quickly**. We know that there have been several **ice ages**, for example.
- a single **catastrophic** event, such as a **volcanic eruption** or a **meteor/asteroid strike**, alters the environment. For example, an ash cloud can block out the light, preventing photosynthesis, and leave a layer of sediment in the sea.

It is important to appreciate that climate change is continual, and was not always man made. In recent times, humans have often been the cause of extinction, but we cannot be blamed for most extinctions going back 2.3 billion years.

The formation of new species is **progressive**. Evolution drives populations to change and adapt. Species have their day, flourish, and then get **out-competed** by better-adapted organisms.

The very earliest fossils may have been **destroyed** by geological activity. Many fossils may still be **undiscovered**, or perhaps we have only **part of the skeleton**. This means that there are **gaps in the fossil record**. But the fossil record shows without doubt that **organisms have changed over time**. This is evidence for **evolution**.

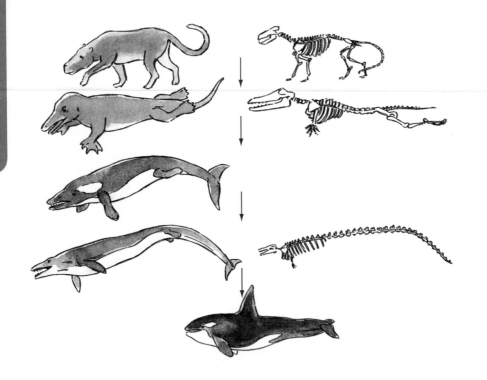

← By looking at and comparing skeletons, scientists can often work out the story of how an animal evolved. The fossil record shows that the whale may have evolved from a land-living animal

What is a species?

Revised

The definition of a species is a population (or populations) of **similar** organisms that can **interbreed** and produce **fertile offspring**.

Fertile means that the offspring are capable of reproduction too. A **horse** can reproduce with a **donkey** to produce an **ass**. But the ass is **not fertile**, so the donkey and the horse are separate species.

How do new species arise?

Revised

New species evolve from ones that already exist. This process is called **speciation**. There are three vital points to appreciate about speciation:

1   There is always **variation** within a population. Organisms differ in many different ways. This is because they possess **different alleles**, or different **combinations of alleles**.

2   Populations can become **isolated**, which means that two or more parts of a population become separated so they **cannot breed** with each other. You cannot have speciation without isolation. The commonest type is **geographical isolation**, where the two populations cannot mix due to some physical barrier, such as a stretch of water.

3   Natural selection will act in **different ways** on the **different populations**. The lucky individuals with the better alleles will have an advantage —birds with longer beaks, for example — so those individuals will have a **better chance** of passing their alleles on to the next generation. Over time, genetic differences will accumulate, so that even if the populations come into contact again they will not be able to reproduce.

Check your understanding

Tested

34 In 1953 a virus was introduced into the wild rabbit population that caused the deadly disease myxomatosis. By 1955, 95% of the rabbit population were dead. Most of the rabbits in the UK today are now immune to the virus.

Describe how this immunity evolved.   *(4 marks)*

Answers online   **Test yourself online**   Online

Dissolved substances

Diffusion

Revised

We have already seen in Biology 2 that **diffusion** is a vital process. It is the movement of substances from regions of high to regions of low concentration until evenly spread. For example, oxygen diffuses into cells and carbon dioxide diffuses out.

The golden rules of diffusion are as follows:

● The greater the **concentration difference** (or **gradient**), the faster the rate.

● The shorter the **diffusing distance**, the faster the rate.

● The greater the **surface area** (between the two areas), the faster the rate.

● The higher the **temperature**, the faster the rate — because molecules have more kinetic energy and move faster.

This is why all organs that have evolved to maximise exchange, such as **lungs**, **gills**, **intestines** and plant **roots**, will have the following key features: large surface area, thin membranes, good/efficient blood supply (if there is one).

In the small intestine, for example, there are

● villi to increase the surface area

● many capillaries — a good blood supply to maintain a diffusion gradient

● thin epithelial cells to minimise the diffusing distance

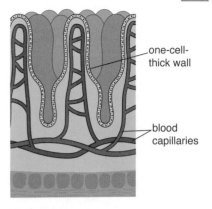

↑ **Actually, these villi look like short hairs about 1 mm long, so that gives an indication of the surface area increase that they create**

one-cell-thick wall

blood capillaries

examiner tip

If you are asked to **explain** in an exam question, and it is for more than 1 mark, you need to put in works like 'so' or 'because'. For example, 'the alveolar epithelium is thin to minimise the diffusing distance, so that diffusion is as rapid as possible'.

Making diffusion even better

Revised

Diffusion is a vital process, but it has two basic limitations:

● It can be too **slow**.

● It **stops** when substances are evenly spread, so you can only ever exchange **half** of the available material.

To overcome this, two processes have evolved:

1 **Faciliated diffusion** is where diffusion in or out of cells is **speeded up** by having special proteins in the cell membrane. Facilitated simply means 'helped'.

2 **Active transport** is where substances can be exchanged beyond equilibrium, i.e. **against a diffusion gradient**.

Active transport is a big advantage to cells because it allows an organism to move materials against a diffusion gradient. In this way an organism can absorb all of a vital substance, such as a trace element or amino acid.

There are three vital points about active transport:

● It takes place **against a concentration gradient**.

● It uses specific **proteins** in the cell membrane.

● It needs **energy**, provided by **respiration** of the cell.

→ **Can you explain in detail what is being shown in these two diagrams?**

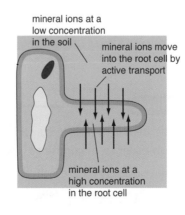

mineral ions at a low concentration in the soil

mineral ions move into the root cell by active transport

mineral ions at a high concentration in the root cell

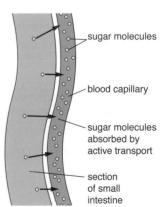

sugar molecules

blood capillary

sugar molecules absorbed by active transport

section of small intestine

Osmosis is the diffusion of water

If you put a piece of plant or animal tissue into some pure water, it will absorb water and swell up. If you put a similar piece of tissue into seawater, it will shrivel up. This is due to **osmosis** — the movement of water into or out of cells.

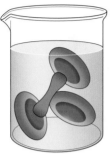

correct concentration
of water
Cells normal

low concentration
of water (brine)
Cells shrivel

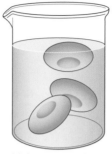

high concentration
of water
Cells swell and burst

← This is what happens to red blood cells in solutions with different concentrations of water

Osmosis is the diffusion of water from a dilute solution to a more concentrated solution through a partially permeable membrane that allows the passage of water molecules.

Cell membranes are permeable to water. All cells contain water and are surrounded by water. All cells therefore face the problem of taking in or losing too much water, and so swelling and bursting or shrivelling up. Animals and plants have different solutions to the problem:

- Plant cells have a rigid **cell wall** that stops cells from swelling and bursting.
- Animal cells do not have cell walls, so animals must control how much water is absorbed into, or excreted from, their bodies — this is **osmoregulation**. In humans the **kidneys** are the main organs of osmoregulation (see pages 76–77).

    → When two solutions are separated by a partially permeable membrane, osmosis occurs. There is a net movement of water molecules into the solution with the highest concentration of solute (dissolved substance), such as glucose

solute molecule   water molecule

examiner tip

Osmosis can be explained in three words: solute attracts water.

Isotonic sports drinks

When we exercise our muscles we use glucose as a source of energy, and we sweat to keep cool. So when we are active we lose glucose, salt and water. If they are not replaced, the ion and water balance of the body is disturbed and the cells do not work as efficiently.

For example, muscles, including the heart, need potassium ions to function correctly. Isotonic drinks have the **same concentration** of dissolved substances as our blood and body fluids.

Check your understanding

1 The young of frogs and toads are called tadpoles and they have gills. Predict what features these gills will have in order to maximise gas exchange. *(3 marks)*

2 Explain why we cannot drink seawater to cure dehydration. *(2 marks)*

Answers online — **Test yourself online** —

Gaseous exchange

Why exchange gas?

- All living things must **respire** — this is the release of energy from organic molecules such as glucose.
- Respiration **uses oxygen**, and **produces carbon dioxide**.
- So all living things must **exchange** these gases.
- Large organisms — and especially those with high energy demands, such as mammals — need to exchange lots of gas.
- That is why special gas exchange organs such as **lungs** and **gills** have evolved.

Lungs

The purpose of the lungs is to get as much air as possible in contact with the blood.

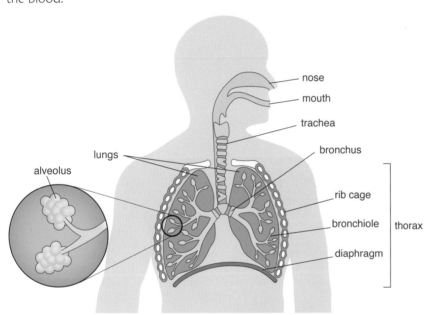

← The lungs are in the upper part of the body (thorax) and are protected by the ribcage. The diaphragm separates the thorax from the abdomen

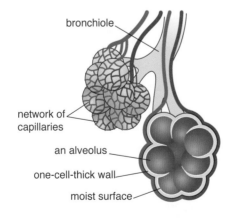

← A cluster of alveoli showing the specialised features for gaseous exchange

To provide a large surface area, millions of alveoli are surrounded by a dense network of blood capillaries. Both the alveolar epithelium and the capillary wall (called the endothelium) are very **thin**, so the distance between air and blood is tiny and **diffusion** of gases in and out is very **rapid**.

So why do we breathe?

Revised

The correct term for breathing is **ventilation**. If we did not breathe, diffusion of gases between the air and blood in the alveoli would soon come to a stop because there would be an **equilibrium**. This is bad. The body becomes starved of oxygen and carbon dioxide builds up.

But, if we keep breathing, there is always fresh air in the lungs and so there is always more oxygen in the air than in the blood. This is another example of **maintaining a diffusion gradient** and it comes up time and time again in biology.

The same idea applies to blood flow through the lungs. If blood flow stopped, equilibrium would be reached and diffusion would stop. So the idea of a **good blood supply** is an important one too. Again, the reason for a good blood supply is that it maintains a diffusion gradient.

> ### examiner tip
> Many candidates state that 'alveoli are moist', as if it is an adaptation, like thin membranes. All gas exchange surfaces are moist because they are permeable to small molecules, so water will always pass out of the cells.

How do we breathe?

Revised

Lung tissue has no muscle, just alveoli, blood vessels and some other cells to hold the whole thing together. However, the lungs are attached to the **ribs** and **diaphragm**, which do have muscle.

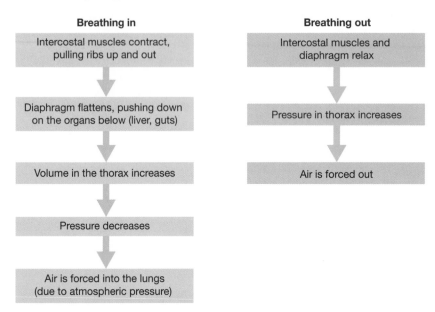

Breathing in

Intercostal muscles contract, pulling ribs up and out

↓

Diaphragm flattens, pushing down on the organs below (liver, guts)

↓

Volume in the thorax increases

↓

Pressure decreases

↓

Air is forced into the lungs (due to atmospheric pressure)

Breathing out

Intercostal muscles and diaphragm relax

↓

Pressure in thorax increases

↓

Air is forced out

Check your understanding

Tested

3   The lungs contain no muscle. Explain how they inflate.   *(5 marks)*

4   **a)** Describe how our breathing pattern changes when we exercise.

(2 marks)

b) Explain the need for this change.   *(3 marks)*

Answers online   **Test yourself online**   Online

Exchange systems in plants

What do plants need to exchange?

- Plants photosynthesise when it is **light**, but **respire all the time**.
- In the daytime, the rate of photosynthesis is faster than the rate of respiration, so the leaves absorb carbon dioxide and give out oxygen.
- The carbon dioxide **diffuses** from the air, through the **stomata** and into the palisade cells. In the daytime the stomata are open so as much CO_2 as possible can get in.
- The plant also needs a supply of **water** and **mineral ions**.

What do the roots do?

- As well as anchoring the plant and providing storage, plant roots have the vital role of absorbing **water** and **mineral ions** from the soil.
- The surface area of the root is greatly increased by **root hairs**, which are specialised epithelial cells.
- Water is absorbed through the roots by **osmosis**.
- Mineral ions such as **nitrate** and **phosphate** are absorbed by **active transport**.

> ### examiner tip
> Many students state that roots absorb food, or carbon, or carbon dioxide. They do not.

Transpiration

Plants absorb a lot of water through their roots. The water passes up the stem to the leaves, where it evaporates from the wet cell walls of the mesophyll cells in the leaves and is lost to the atmosphere through the stomata.

This loss of water from the leaves is called **transpiration**. The passage of water through the plant is known as the **transpiration stream**.

Transpiration is basically evaporation, so it will be greatest when conditions are **warm**, **windy** and **dry**.

If the plant loses more water by evaporation than it gains through the roots, it will begin to wilt.

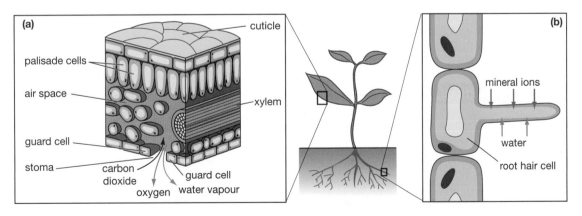

↑ a) A cross-section of a leaf showing the movement of materials through the stomata. What process is shown here? b) Root hair cells give the roots a large surface area for absorption, and internal air spaces give leaves a large surface area for gas exchange

● The word **stoma** simply means 'hole' (**stomata** is plural — lots of holes). Stomata are created by two banana-shaped **guard cells**.

● If the plant is losing too much water, guard cells will also lose water, change shape and close the opening. This reduces water loss by transpiration.

● To open the stomata again, the guard cells absorb water, change shape to become more curved, so widening the stomatal opening.

● At night there is no photosynthesis, so the stomata close to reduce water loss.

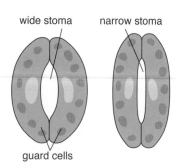

wide stoma   narrow stoma

guard cells

↑ Stomata become smaller if a plant needs to reduce the amount of water being lost by transpiration

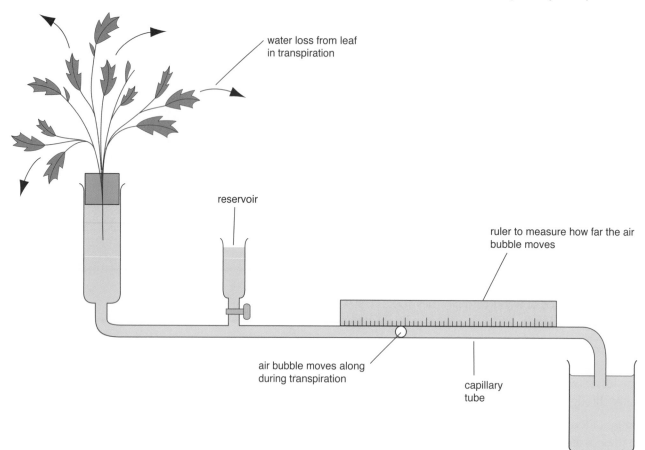

water loss from leaf in transpiration

reservoir

ruler to measure how far the air bubble moves

air bubble moves along during transpiration

capillary tube

↑ The potometer is a device that measures the rate of water loss by transpiration. It is basically a transparent extension of the xylem. As the water evaporates from the leaves, a column of water is pulled up the xylem, and its progress can be followed by watching the air bubble. This can be used to compare the rate of transpiration in different plants, or to investigate the effect of temperature, humidity and wind

Check your understanding ─────────────────────────── Tested

5 Define the term transpiration.   (1 mark)

6 Under what conditions is transpiration slowest?   (2 marks)

7 Plants need nitrogen in order to make proteins. Explain how plants obtain their nitrogen.   (2 marks)

Answers online   **Test yourself online** ──────────────── Online

The blood system

Why do we need a circulation?

Revised

Blood is a **mass flow** or a **mass transport** system. Large volumes of fluid are pumped rapidly around the body from organs of exchange such as the guts and lungs, to the cells that need them.

The heart is simply a pump.

Mammals have a double circulation — there are two circuits that take blood on a return journey to the heart.

1 The **pulmonary circulation** carries blood to the lungs and back.

2 The **systemic circulation** carries blood to the rest of the body and back.

We need two circulations because blood goes to the lungs to collect oxygen, but in doing so it **loses pressure**, so it needs to return to the heart for a **boost**. In order to pump blood round two circulations at the same time, we need a four-chambered heart:

● **Two atria**. Their job is simply to fill with the right volume of blood, and pump it into the ventricles. Think of them as 'loading chambers'.

● **Two ventricles**, whose job is to create **pressure**. They contract powerfully, forcing blood into arteries.

The **right** side of the heart fills with **deoxygenated** blood, and pumps blood around the pulmonary circulation.

The **left** side of the heart fills with **oxygenated** blood from the lungs, which needs to be pumped around the systemic circulation. This requires more pressure, so the left ventricle has a thicker muscle wall than the right ventricle.

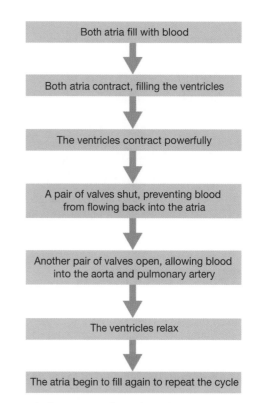

↑ The events of one heartbeat

[Flowchart contents:]
Both atria fill with blood
↓
Both atria contract, filling the ventricles
↓
The ventricles contract powerfully
↓
A pair of valves shut, preventing blood from flowing back into the atria
↓
Another pair of valves open, allowing blood into the aorta and pulmonary artery
↓
The ventricles relax
↓
The atria begin to fill again to repeat the cycle

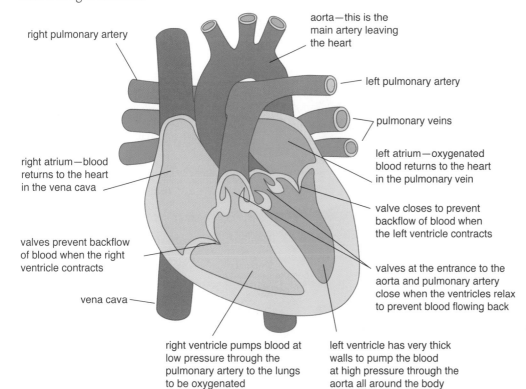

← Diagram of a section through the heart. Note the positions of the valves and the wall thickness

right pulmonary artery

aorta—this is the main artery leaving the heart

left pulmonary artery

pulmonary veins

right atrium—blood returns to the heart in the vena cava

left atrium—oxygenated blood returns to the heart in the pulmonary vein

valve closes to prevent backflow of blood when the left ventricle contracts

valves prevent backflow of blood when the right ventricle contracts

valves at the entrance to the aorta and pulmonary artery close when the ventricles relax to prevent blood flowing back

vena cava

right ventricle pumps blood at low pressure through the pulmonary artery to the lungs to be oxygenated

left ventricle has very thick walls to pump the blood at high pressure through the aorta all around the body

Blood vessel walls reflect their function

There are three main types of blood vessels: **arteries**, **veins** and **capillaries**.

1 Arteries carry blood away from the heart. They have thick, muscular, **elastic** walls in order to cope with the high **pressure** created when the heart beats.

2 Veins return blood to the heart. They have thinner walls because they do not have to cope with high pressure. They have **valves** to prevent blood flowing backwards.

3 Capillaries are tiny blood vessels whose walls are **one cell thick**. The walls are **permeable** because their function is to allow exchange of materials between blood and the surrounding tissues.

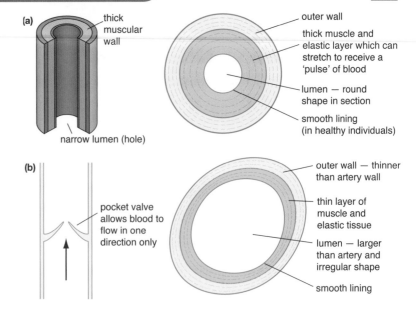

(a)
- thick muscular wall
- narrow lumen (hole)
- outer wall
- thick muscle and elastic layer which can stretch to receive a 'pulse' of blood
- lumen — round shape in section
- smooth lining (in healthy individuals)

(b)
- pocket valve allows blood to flow in one direction only
- outer wall — thinner than artery wall
- thin layer of muscle and elastic tissue
- lumen — larger than artery and irregular shape
- smooth lining

↑ a) Note the thickness of the artery wall and the shape of the transverse section. b) Note the irregular shape and the thinner muscle and elastic layer of the vein

Heart problems

Heart disease is the commonest cause of death in the western world. Many things can go wrong with the heart, but by far the commonest problem is **coronary heart disease**, in which the arteries become narrower due to a build-up of **fatty deposits** inside them. When the **coronary arteries** get blocked, not enough oxygen and glucose reaches the heart muscle, part of the heart muscle dies, and a **heart attack** results.

When coronary arteries narrow, they can be treated by inserting a **stent** — a tubular wire mesh that keeps the lumen of the artery open.

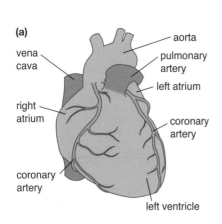

(a)
- vena cava
- right atrium
- coronary artery
- aorta
- pulmonary artery
- left atrium
- coronary artery
- left ventricle

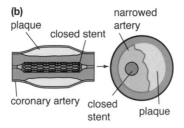

(b)
- plaque
- closed stent
- narrowed artery
- coronary artery
- closed stent
- plaque

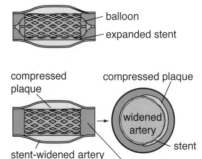

- balloon
- expanded stent
- compressed plaque
- compressed plaque
- stent-widened artery
- widened artery
- stent
- increased blood flow

← a) The coronary arteries serve the heart muscle tissue. b) This stent allows blood to flow freely again

Blood flows from the heart into arteries, then into smaller arteries, which finally branch out into capillaries. These tiny vessels take blood to within a fraction of a millimetre of all the respiring cells.

All cells are surrounded by **tissue fluid**, from which cells get the oxygen and nutrients, and into which they secrete their waste and other products.

Blood slows down as it flows along a capillary. The walls are one cell thick and permeable (leaky) so that exchange can happen between blood and tissue fluid. Tissue fluid is basically plasma that leaks out of the capillaries.

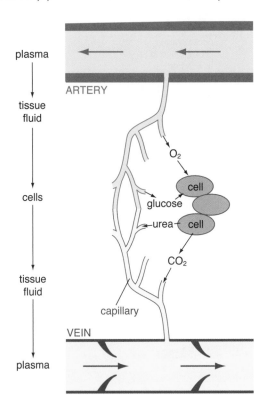

← Tissue fluid forms at the artery end of a capillary, and drains back towards the vein end

8 For each of the following blood vessels, state the pressure and the oxygen content of the blood it contains (you could do this as a table):

  a) vena cava   (1 mark)

  b) pulmonary artery   (1 mark)

  c) pulmonary vein   (1 mark)

  d) aorta   (1 mark)

9 Blood flow is slowest in capillaries. Suggest an advantage of this slow flow.
                                                (1 mark)

10 Arteries carry oxygenated blood. Is this always true? Explain.   (1 mark)

11 Most arteries lie deep under the surface of the skin, while veins run much closer to the surface. Explain the advantage of this arrangement.   (2 marks)

Answers online — Test yourself online Online

The blood

Blood is a complex fluid containing **red blood cells**, **white blood cells**, **platelets** and **plasma**.

Red blood cells are basically little bags of **haemoglobin**. They have no nucleus or any of the other normal cell contents such as mitochondria. The function of haemoglobin is to collect oxygen where it is abundant — the **lungs** — and release it where it is needed — the **respiring tissues**.

As red blood cells pass through the lungs, haemoglobin combines with oxygen to form bright red **oxyhaemoglobin**. As it passes through the respiring tissues in the rest of the body, the haemoglobin gives up its oxygen, becoming a darker red as it does so.

White blood cells form a vital part of the **immune system**, which defends the body against pathogens (this was covered in Biology 1).

Platelets are fragments of cells. They too have no nucleus but are smaller than red cells. They contain various blood clotting factors that help blood to clot when a vessel has been broken.

Plasma is the fluid part of blood. It is a solution containing:

● the **products of digestion**, such as glucose, amino acids, fatty acids and glycerol, vitamins and mineral ions

● **waste** products. Amino acids are broken down in the **liver** to make **urea**. The kidneys extract the urea, which we excrete in the **urine**.

● **carbon dioxide**, mainly carried as hydrogencarbonate ions HCO_3^-

Plasma also contains other important substances such as **hormones**, **blood clotting** factors and **heat**.

> **examiner tip**
> If you need to list the contents of plasma, think about the components of a balanced diet, and what they are absorbed as.

> **examiner tip**
> Blood is defined as a tissue, like cartilage and bone.

Check your understanding
Tested

12 Red blood cells do not live very long. Use your knowledge of their structure to suggest why. *(2 marks)*

13 The higher the altitude at which people live, the greater the volume of red blood cells in their blood. Suggest an advantage of this adaptation. *(2 marks)*

14 Blood clotting is a relatively fast reaction. Give two advantages of this fact. *(2 marks)*

Answers online   **Test yourself online**
Online

Transport systems in plants

Revised

examiner tip

There are several different types of plant, including mosses, ferns and conifers. However, the most successful group are the flowering plants. All plant topics covered in this book refer to flowering plants.

Flowering plants have a transport system that consists of two different tissues: **xylem** and **phloem**. Both types of tissue are made from long, cylindrical cells that join up at the ends to form a continuous conducting pathway. Think of a bundle of drinking straws joined up end to end.

Xylem

The function of xylem is to transport water and dissolved mineral ions from the **roots**, up the **stem** to the **leaves** and other organs, such as flowers and fruits.

Water enters the root hair cells by **osmosis**, and moves through the root tissue to the xylem by the same process.

Once in the xylem, water is drawn up the plant due to evaporation from the leaves, i.e. **transpiration**. As water evaporates, it creates a tension in the xylem that draws water upwards, even to the top of the tallest trees. This is the driving force behind the movement of water through a plant.

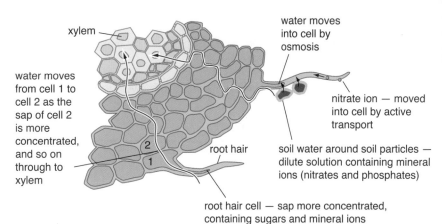

↑ **Active transport and osmosis in a root**

Phloem

The function of phloem tissue is to transport **dissolved sugars** (usually **sucrose**) from where they are made — the leaves — to where they are needed — the rest of the plant. In particular:

● **Growing points** where new cells are made
● **Storage organs**, often in the roots (think of potatoes, carrots and turnips)

Movement of dissolved sugars in the phloem is called **translocation**. The direction of movement in the phloem can be either **up or down** the stem, depending on where the sugars are needed.

examiner tip

Remember the Fs: food flows in phloem. This is not a good thing to write in an exam, but it will stop you from confusing it with xylem.

Check your understanding

Tested

15 Phloem tissue transports dissolved sugars. Where did these sugars come from? *(1 mark)*

16 Give two similarities and two differences between xylem and phloem. *(4 marks)*

Answers online   Test yourself online   Online

Removal of waste and water control

Kidneys

Kidneys have three functions:

1 **excretion**

2 control of **water balance**

3 control of **ion balance** (think of it as 'salt balance')

Excretion is the removal of **waste**, which is defined as the **by-products of metabolism**. Waste products that have to be removed include:

● **carbon dioxide**, produced by respiration and removed via the lungs when we breathe out

● **urea**, produced in the liver by the breakdown of excess **amino acids** and removed by the kidneys in the urine, which is stored in the bladder before leaving the body

Our kidneys remove all of the urea, but will **excrete** or **conserve** water and salt depending on whether we have too much or too little.

1 If there is not enough water, or too many ions, blood and body fluids will be **too concentrated**. Water will be drawn out of cells by **osmosis**, making cells shrivel.

2 If there is too much water or too few ions, the blood and body fluids will be **too dilute**. Water will enter cells by osmosis, making them swell up.

How the kidneys work

Considering their size, the kidneys receive a lot of blood. Kidneys function by first **filtering** the blood and then **reabsorbing** what the body needs. When blood is filtered, a clear liquid similar to **tissue fluid** is produced. This contains lots of useful substances. The kidney will then:

● reabsorb all the **glucose** — an example of active transport

● reabsorb the **dissolved ions** needed by the body

● reabsorb as much **water** as the body needs

What is left is a solution of urea and excess ions. This is urine.

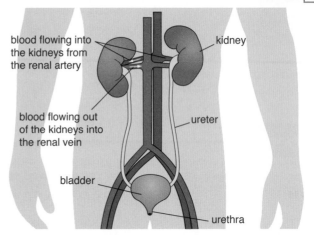

blood flowing into the kidneys from the renal artery

blood flowing out of the kidneys into the renal vein

kidney

ureter

bladder

urethra

↑ **The kidneys are attached to the circulatory system by the renal arteries and renal veins. The urine that is produced is transported to the bladder via the ureter. How does the composition of the blood in the renal vein differ from that in the renal artery?**

When kidneys fail

When kidneys fail, there are three big problems:

● accumulation of urea

● no control of water levels

● no control of salt levels

People suffering from failure in both kidneys are in urgent need of treatment. To some extent, the problems can be minimised by a **diet** in which there is strict control of protein, water and salt intake. But, diet will not solve the problem.

The two effective treatments are:

1 kidney dialysis

2 kidney transplant

Dialysis machines are basically artificial kidneys that take blood out of the body and filter it against a special **dialysing membrane**. Dialysis is a physical process that relies on **diffusion** to restore normal urea, water and ion levels in the blood.

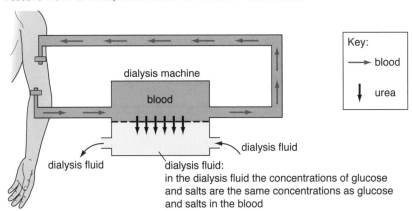

Key:

⟶ blood

↓ urea

← A dialysis machine removes urea from the blood without removing useful substances

dialysis machine

blood

dialysis fluid

dialysis fluid

dialysis fluid: in the dialysis fluid the concentrations of glucose and salts are the same concentrations as glucose and salts in the blood

In a dialysis machine a person's blood flows between **partially permeable membranes**. The question is, how does the machine take out all the urea without losing all the valuable substances such as glucose? The key to the machine's success is the dialysis fluid, which contains the same concentration of useful substances as the blood, but no urea. Diffusion of glucose and useful mineral ions will be **equal in both directions**, so none is lost. Urea, however, **only diffuses out** from the blood into the dialysis fluid.

Kidney transplants

Revised

In kidney transplants a diseased kidney is replaced with a healthy one from a **donor**. The big problem with transplants from different people is that the organ will have **different proteins** on its cells. These act as **antigens** and allow the immune system to recognise the donor organ as foreign. The kidney may be **rejected** by the immune system unless precautions are taken.

To prevent rejection, a donor kidney with a **tissue-type** similar to that of the recipient is used. This means that the **proteins** on the donor and recipients cells are as **similar as possible**.

However, unless the donor and recipient are identical twins, there will never be a perfect match. To prevent rejection the recipient is treated with drugs that suppress the immune system. There are some very effective **immunosuppressant** drugs available, but they leave the individual **immunocompromised** to some extent and less able to fight off disease.

examiner tip

Many people confuse urea with urine. Urea is a compound — a white solid. Urine is a solution containing varying amounts of salt and urea.

When describing what the kidneys do, it is too vague to talk of 'separating the good stuff from the bad stuff'. The kidneys filter the blood according to the needs of the body.

Check your understanding

Tested

17 Suggest how the volume and concentration of your urine will be affected if you:

   a) drink lots of water   *(1 mark)*

   b) exercise, sweat, and do not drink   *(1 mark)*

18 Why is it an advantage to reabsorb glucose by active transport and not diffusion?   *(2 marks)*

19 When blood is filtered, the resulting liquid is clear. Explain why.   *(1 mark)*

20 Which of the following will be found in the urine of a healthy person?   *(2 marks)*

   water   glucose   salt   urea   blood   protein

Answers online   **Test yourself online**   Online

Temperature control

Humans, like all mammals (and birds), can maintain a **stable core temperature** at about 36–37°C. The core is the centre of the body — the head and trunk. Our hands and feet may get cold, but the core is remarkably constant, except when we have a fever.

Our bodies **generate heat** as a by-product of processes such as **respiration** and **muscular contraction**. Generally, we keep our core temperature constant by **maximising** or **minimising heat loss** to the surroundings.

● Body temperature is monitored and controlled by the **thermoregulatory centre** in the brain.

● This centre has **receptors** sensitive to the temperature of blood flowing through the **brain**.

● There are also temperature receptors in the **skin**.

● All these receptors send **nerve impulses** to the thermoregulatory centre, giving information about body temperature. The thermoregulatory centre sends out nerve impulses to other parts of the body to bring about increases or decreases in heat loss.

● Sometimes a change in our **behaviour** is all that is needed to stay the right temperature. Our skin can act as an early warning, and we can, for example, put on extra clothing or open a window.

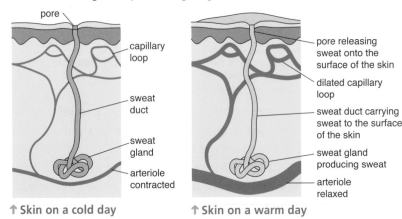

↑ **Skin on a cold day**　　↑ **Skin on a warm day**

When the core temperature drops
Revised

● **Shivering**: rapid contraction and relaxation of the muscles creates heat (shivering muscles **respire** more quickly, and heat is a by-product).

● **Vasoconstriction**: small blood vessels leading to the skin get smaller (**constrict**) so less blood flows to the surface of the skin. This keeps the warm blood deeper in the body, and less heat is lost by radiation from the skin.

● **Hairs stand on end**: to trap a layer of insulating air.

When the core temperature rises
Revised

● **Vasodilation**: blood vessels leading to the surface of the skin get larger (**dilate**) so that more blood can flow to the surface. Heat from the blood is transferred to the skin and sweat.

● **Sweating**: sweat glands secrete a salty solution onto the skin. The process of **evaporation** has a cooling effect. It causes water loss, which has to be replaced by drinking.

Check your understanding
Tested

21 a) Where in the body is the temperature control centre? *(1 mark)*

  b) Explain how the body responds to a drop in core temperature.
  (4 marks)

22 a) State the relationship between internal and external temperature for type A and type B organisms in the diagram.
  (2 marks)

  b) Which of the following would be examples of species A? *(1 mark)*

  lizard　goldfish　crocodile　chicken　frog　dog

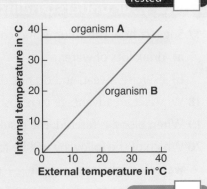

Answers online　Test yourself online　Online

Sugar control

The food we eat contains a lot of glucose. It is common in sweet foods, and even more is locked up in starchy foods such as bread, wheat, rice and potatoes. The starch is digested into glucose and absorbed into the blood from the small intestine.

- Glucose is our main fuel — cells use it in **respiration**.
- The glucose level in the blood must stay within **certain limits**.
- If there is too much glucose, we take it into our cells and store it as **glycogen**, which is a bit like starch — lots of glucose molecules in a chain.
- If there is too little glucose, we break down our glycogen reserves and release more into the blood.

Controlling blood sugar levels
Revised

- The **pancreas** is the key organ responsible for monitoring and controlling blood glucose.
- The **liver** is also important because it receives all of the glucose from the gut, and stores a lot of it as glycogen.
- If blood glucose levels are **too high**, the pancreas will detect the change and respond by releasing **insulin** into the blood. This **hormone** — like all hormones — travels to all parts of the body, and works by allowing cells to **take up more glucose**. So, glucose enters cells and the blood glucose levels get lower.
- If blood glucose levels are too low the pancreas also detects the change and releases another hormone, **glucagon**, into the blood. **Glucagon** causes the breakdown of glycogen into glucose.

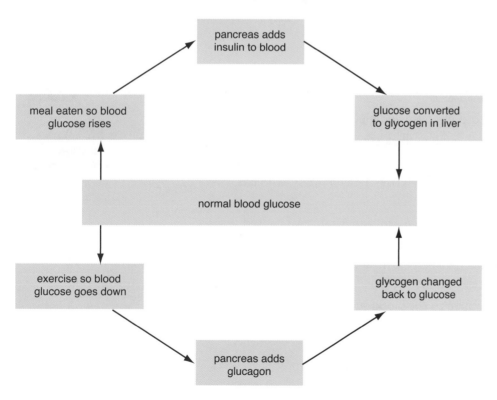

↑ **Two hormones work to control blood glucose level**

Diabetes

Diabetes is a very common disease. It is a **failure** to **control blood glucose levels**. If blood glucose levels are too high, it can be fatal.

- In **type 1 diabetes**, the pancreas fails to make enough, if any, insulin. People with diabetes need to **test their blood sugar** frequently and have regular **insulin injections**.

- In **type 2 diabetes**, also called **late-onset diabetes**, the body makes some insulin, but for one reason or another the body is not able to control blood sugar levels effectively. This type of diabetes is associated with being overweight and can sometimes be treated by carefully controlling **diet** and **exercise** levels.

> **examiner tip**
>
> There are many more type 2 diabetics than type 1, but it is type 1 you will be asked about in the exam.

Since the discovery of insulin, people with diabetes have had a normal life expectancy. Monitoring blood glucose levels, having injections and watching what you eat are inconvenient, but have saved many lives.

To replace insulin injections, new methods of getting insulin into the body are being developed. These include skin patches and continuous-release insulin pumps. Future treatments could include insulin inhalers, and transplants of pancreatic cells or stem cells so the pancreas starts to produce insulin again.

> **examiner tip**
>
> You do not need to remember the new treatments for diabetes, but you do need to be able to describe their benefits for people with diabetes, such as improvement in quality of life, less risk of infection from dirty needles and fewer 'highs and lows' in blood sugar levels.

Check your understanding

23 Distinguish between glucose, glycogen and glucagon. *(3 marks)*

24 Suggest two things that can happen to glucose once it has been absorbed into a cell. *(2 marks)*

25 The table shows the blood sugar levels over 1 day of Mary, who has diabetes.

a) Suggest what happened between 6 a.m. and 8 a.m. that caused the blood sugar levels to rise. *(1 mark)*

b) Suggest what happened between 2 p.m. and 4 p.m. that caused the blood sugar levels to decrease. *(1 mark)*

c) Predict what might happen to the blood sugar levels of someone with diabetes if they exercised more than usual. *(1 mark)*

d) Predict what might happen to the blood sugar levels of someone with diabetes if they took too much insulin. *(1 mark)*

Time	Blood sugar (arbitrary units)
6.00 a.m.	4
8.00 a.m.	10
10.00 a.m.	12
12.00 noon	13
2.00 p.m.	18
4.00 p.m.	10
6.00 p.m.	11
8.00 p.m.	8

Answers online   **Test yourself online**

Waste from human activity

There are almost 7 billion people on this planet. That is seven thousand million. Although the populations of richer countries, such as the UK, are relatively stable, in many countries the birth rate exceeds the death rate, and so the population is expanding.

Effect on resources

- A rising population needs more food, fuel and clean water.
- Raw materials, including **non-renewable energy** resources, are rapidly being used up.
- **Deforestation** is not just about using timber, but replacing rainforest with land used to grow crops to produce more food.
- People want a higher **standard of living**, and more resources are used to make more cars, televisions or tumble dryers, or to travel to far-away places.
- Expanding cities and more farms, landfill sites and quarries reduce the amount of land available for animals and plants.

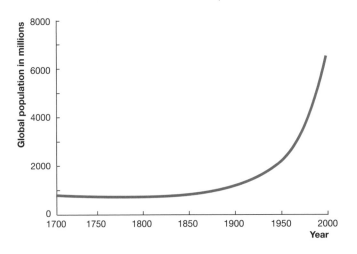

← Since about 1800, improvements in medicine and food production have meant that many more people survive, and so go on to reproduce

Production of waste

Extracting metals, quarrying rocks and drilling for oil are all processes that produce large amounts of waste. We also dump vast quantities of household refuse and sewage into the land and sea. When human activity upsets the natural balance of ecosystems, it is known as **pollution**.

Pollution can kill organisms directly or affect or destroy **habitats** for wildlife, which reduces **biodiversity**.

Air pollution
- Carbon dioxide from **burning of fossil fuels** is thought to be causing **global warming** (see page 85).
- **Smoke** and **soot** pollute the air we breathe.
- **Sulfur dioxide** from burning of fossil fuels can combine with water vapour in clouds to produce **acid rain**, which pollutes lakes and streams.

Land pollution

● **Pesticides** and **herbicides** (weedkillers) sprayed onto farmland can affect populations of wild birds, insects and flowers.

● **Toxic chemicals** from industrial waste, such as **heavy metals**, may be washed from land into water.

Water pollution

● **Sewage** from farmland and water-treatment works can over-fertilise lakes and waterways and reduce the oxygen available for aquatic creatures (bacteria take all the oxygen).

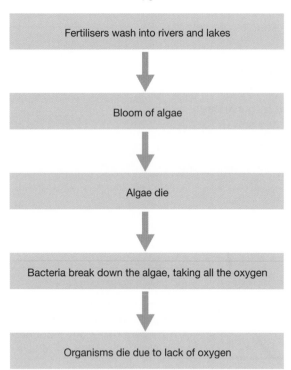

Fertilisers wash into rivers and lakes

↓

Bloom of algae

↓

Algae die

↓

Bacteria break down the algae, taking all the oxygen

↓

Organisms die due to lack of oxygen

← **Fertilisers and sewage are not poisonous as such, but they cause problems by over-fertilisation. This upsets the aquatic ecosystem**

● **Toxic waste** from industry — even tiny amounts — can kill species. Examples of toxic waste include **mercury**- and **lead**-based chemicals.

● **Pesticides** and **herbicides** can get washed off farmland into waterways. At low doses these are often harmless, but they can be absorbed by simple organisms such as algae. Often, these chemicals cannot be excreted so they **accumulate** up the food chain. The higher up the food chain, the greater the dose in each animal. Eventually a dose is reached where the animal **dies** or **cannot reproduce**.

Check your understanding
Tested

26 Use the following food chain to explain how pesticides can upset the balance of ecosystems. *(4 marks)*

algae → water fleas → stickleback → perch → heron

27 a) Suggest what the term biodegradable means. *(1 mark)*

b) What gases will be produced as a consequence of biodegradation? *(2 marks)*

c) What problems will these gases cause? *(1 mark)*

Answers online ——— Test yourself online ——— Online

Deforestation and destruction of peat areas

Deforestation

Revised

Of all human activity, large-scale **deforestation** in tropical areas is seen as one of the most destructive. It is done to provide timber and to provide land for agriculture, but there are many side effects:

- There is increased release of **carbon dioxide** into the **atmosphere**, due to burning and the activities of microorganisms.

- There is a reduction in the rate at which carbon dioxide is removed from the atmosphere by **photosynthesis.**

- Trees return a lot of water vapour to the atmosphere by **transpiration**. Without the trees, **rainfall patterns change**.

- **Destroyed habitats** for many organisms leads to a **reduction in biodiversity**.

- Many organisms that may have been of **medicinal value** are lost.

- **Isolation** of animal populations occurs because many forest species will not cross roads and open farmland. They stay in small pockets of forest, which leads to **inbreeding**.

- Roots hold the **soil** in place, especially on hillsides. When the trees are gone, rain washes vast amounts of soil into waterways, causing them to silt up.

- With the thick topsoil gone, land will not support crops or cattle, and becomes useless.

Forests act as **reservoirs** of carbon. Wood is a tissue that 'locks up' carbon for decades. Deforestation, especially followed by burning, releases vast amount of carbon dioxide into the environment.

> **examiner tip**
> When things rot, it is because microbes — bacteria and fungi — are breaking them down. The respiration of these organisms releases carbon dioxide into the atmosphere.

The land created by deforestation is used for:

- crops from which **biofuels** can be produced. The plant material is **fermented** to produce **ethanol**, which can be used in engines.

- grazing land for cattle, which produce **methane** (see page 85)

- growing rice fields to provide more food. Rice grows in waterlogged fields that are anaerobic (oxygen poor) and this also produces methane.

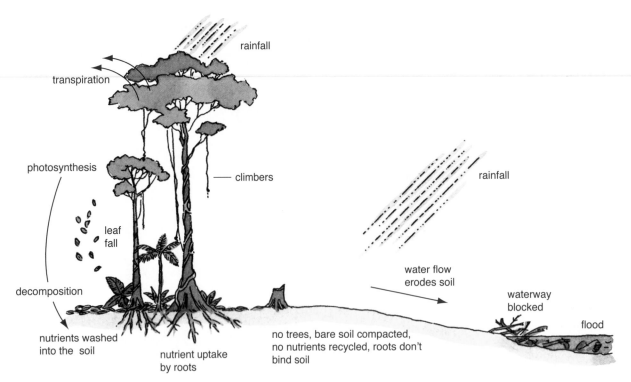

↑ **The rainforest cycle and biological consequences of disturbance**

Labels in diagram: rainfall, transpiration, photosynthesis, climbers, leaf fall, decomposition, nutrients washed into the soil, nutrient uptake by roots, no trees, bare soil compacted, no nutrients recycled, roots don't bind soil, rainfall, water flow erodes soil, waterway blocked, flood

Destruction of peat bogs — Revised

Peat is mainly composed of partly decomposed plant material. Conditions in a peat bog are **waterlogged**, **anaerobic** and **acidic**, with the result that generations of plants die but do not rot completely.

Peat bogs consist of layers upon layers of dead plant material that locks up huge amounts of carbon. Digging up the peat and burning it for fuel releases the carbon back into the atmosphere as carbon dioxide.

There are vast areas of peat bog around the world — an estimated 2% of the total land —though there is very little left in the UK.

As well as being used for fuel, peat is in big demand from gardeners as compost for plants. In an attempt to reduce the amount of peat being taken, 'peat-free' composts are now being marketed.

Check your understanding — Tested

28 Farmers are being encouraged to leave thin strips of trees across areas of farmland, so that areas of forest join up. Explain the benefits of these 'corridors'. *(2 marks)*

29 Explain the effect of deforestation on carbon dioxide concentrations in the atmosphere. *(2 marks)*

30 A UK scientist goes to Brazil to try to persuade farmers to stop deforestation. The Brazilian farmer says, 'The UK cut all its forests down a long time ago. This is the only way we can feed our children. Go away.' Suggest what you might say to make him change his mind. *(3 marks)*

Answers online — **Test yourself online** — Online

Biofuels and global warming

The Earth's atmosphere has been getting warmer over the past 200 years. Scientists think that this is due to an increased **greenhouse effect**.

What causes the greenhouse effect? — Revised

As in a greenhouse, the atmosphere allows **sunlight energy in**, but does not let all of it escape, so it gets hotter.

1 Temperatures on Earth result from a **balance** in **energy received** and **energy lost**.

2 The energy we get from the Sun is not changing, but we are **losing less**.

3 This is because we are **changing** the **composition** of our **atmosphere**. Gases such as **carbon dioxide** and **methane** act together to prevent heat from escaping.

Where does carbon dioxide come from? — Revised

Power stations, industry, cars and aircraft engines all burn **fossil fuels** to produce millions of tonnes of carbon dioxide every year. While there is a massive amount of **combustion** going on, **photosynthesis still absorbs carbon dioxide** from the atmosphere.

Cutting down forests (**deforestation**) is another problem. A lot of carbon is locked up in wood, so when we cut down and burn trees this is released as carbon dioxide back into the atmosphere.

Methane gas (CH_4) is another **greenhouse gas**. Production of methane is increasing due to:

● cows belching and farting (**flatulence**) — cows release large amounts of methane as a by-product of their digestion. As more cattle are being bred, more methane is produced.

● an increase in rice paddy fields to feed the world's growing population. The anaerobic conditions in waterlogged paddy fields generate methane.

Complex effects of global warming — Revised

The Earth's climate is a **complex system**. That means that its effect is impossible to predict exactly. However, here are some possibilities:

● **Glaciers** and **ice caps** at the poles will **melt** and drain into the sea, making **sea levels rise**.

● Higher temperatures make water **expand**, which scientists think may have more effect than melting ice on rising sea levels.

● Rising sea levels will flood low-lying areas with high populations, such as large areas of the Netherlands.

● The extra fresh water may **disrupt ocean currents**. This could be a disaster for the UK — currently kept warmer by Atlantic currents.

● More **severe weather** — for example, there could be more **hurricanes**.

● Higher temperatures could mean more evaporation, more clouds and a change in rainfall patterns.

● All over the world, ecosystems depend on the climate. Changes could disrupt plant growth and patterns of agriculture. Many animals and plants are **inter-dependent**, but react differently to climate change. For example, a particular species of flower may emerge before the insect needed to pollinate it. Birds might migrate at the wrong time, and animals might emerge from hibernation before there is enough food to live on.

Carbon sequestration

Revised

Carbon sequestration is the process of actively removing carbon dioxide from the atmosphere and 'locking it up' in oceans, lakes and forests. Human activities such as deforestation have released billions of tonnes of carbon that was locked up, so a lot of research is being done on trying to reverse the process. Among the many methods being tried are:

- making new peat bogs
- planting new forests
- encouraging growth of algae in oceans (plant plankton)

Biofuels

Revised

Biofuels are fuels that are made by **biological processes** and are therefore **renewable**, as opposed to **fossil fuels**, which are not.

Biogas is a mixture of gases, mainly methane (typically 60% methane, 40% carbon dioxide) that result from the **anaerobic fermentation** of plant material. It is not unlike the gas produced in our intestines when we eat a lot of vegetables.

Biogas is used in developing countries for cooking, heating and generating electricity.

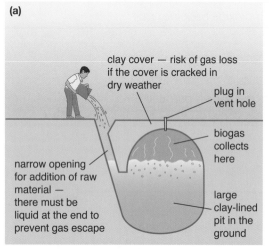
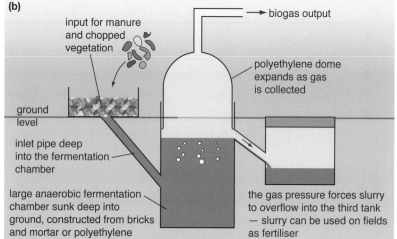

↑ a) **Simple rural biogas generator — this is a batch process. If production rate drops, the tank has to be emptied and cleaned out, and the cover remade. There will be a gap in production. b) Modern biogas generator — this is a semi-continuous process. The manure from two dairy cows can provide a fairly continuous supply of biogas to meet the cooking and lighting needs of a small family**

Check your understanding

Tested

31 Match the words **a)–d)** with the spaces **1–4** in the text below.   (4 marks)

   a) deforestation   **b)** acid rain   **c)** global warming   **d)** sewage

   Burning wood releases sulfur dioxide, which causes _____1_____, and carbon dioxide, which causes _____2_____. The population is increasing, so uses more fuel and produces more _____3_____ as well as causing _____4_____.

32 List two factors contributing to increased methane concentrations in the atmosphere.   (2 marks)

Answers online   Test yourself online   Online

Food production

The problems of energy loss in food chains have a big impact on food production. Look at the food chain that produces beef. There are two links in the chain, so two opportunities to lose energy:

grass → cow → man

The grass contains a lot of indigestible material such as cellulose. The energy in the cellulose passes straight through the animal and is lost in the faeces.

Of the energy the cow does absorb, most of it is lost as heat because the animal is warm blooded.

So, only a small amount of the energy in the grass is converted into meat (increased **biomass** of the cow).

Increasing efficiency of food production

Revised

Since animals need a large amount of energy from food just to move around and keep warm, the efficiency of food production can be improved by keeping food animals in insulated barns.

Limiting the movement of food animals and controlling the temperature of their surroundings increases the efficiency of food production.

↑ In battery farming, the chickens are kept warm and given little space. They use less energy in respiration, and so their energy conversion is more efficient. Many people find this disturbing, but still like cheap food

↑ Cows must maintain a constant body temperature, and so they respire quickly to keep warm, especially in cold weather. It would seem to make sense to farm cold-blooded animals such as crocodiles. They convert more of their food into meat. However, as they are carnivores, you would have to feed them meat in the first place, which in terms of energy makes no sense whatsoever

examiner tip

You may be asked to evaluate the intensive rearing of animals. Remember to give both sides of the argument.

Feed the world?

Revised

To feed more people when resources are limited, it makes more sense for people to eat plants (that is, to be vegetarians, not meat-eaters), as in the food chain: soya → human. There is only one step in the chain, so less energy is lost.

Reducing the number of stages in food chains increases the efficiency of food production.

However, it is not always as simple as choosing to grow crops instead of animals:

● A diet of one or a few crops may not be balanced.

● Many people will want to carry on eating meat.

● Some land is unsuitable for growing crops, but may be suitable for grazing a few goats or sheep.

Energy for food processing and transport

Revised ☐

Even more energy is lost when food is harvested by **machines**, **processed**, **packaged**, **stored** and **distributed** around the world.

A lot of energy is saved if food is consumed where it is produced, to reduce the number of **food miles** it travels. However, so long as people want to eat meat protein such as beef and lamb, we will continue to waste energy producing and distributing it.

Fish stocks are declining

Revised ☐

Many of the world's fisheries have seen a serious decline in stock in recent years. There are more large fishing vessels, many of which use technology to find the fish and large nets to catch more each time.

Fishing must be **sustainable**, so that the more popular food species such as cod have a chance to breed and grow. There are several measures that serve to maintain fish stocks:

● Use nets with a **large mesh size**, so that the smaller fish escape and have a chance to breed.

● Do not fish during the **breeding season**, or in known breeding grounds.

● Have **quotas**, which set limits on the amount of fish that can be caught at any one time.

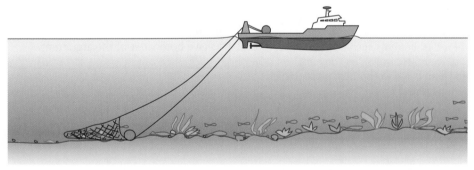

← Bottom trawling for plaice. What damage do the rollers do to the ecosystem?

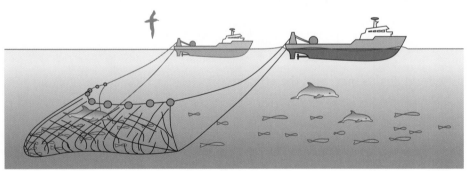

← Pair trawling for cod. What else gets caught in the wide net?

Protein from fungi is called **mycoprotein** and has been popular for a number of years. There are several species of fungi, most notably *Fusarium*, that can be grown quickly and produce protein that is acceptable to vegetarians.

Fusarium will grow quickly if given the right conditions:

● **glucose** syrup, a relatively cheap foodstuff, provides **energy**

● **oxygen** — the fungus respires **aerobically**

● a source of **nitrogen** to make **protein**

The biomass of the fungus increases quickly, after which is it filtered, purified and packaged.

examiner tip

Make sure that you can explain why increased food miles decrease the efficiency of food production.

Check your understanding Tested

33 The diagram represents the energy transfer from grass to a cow. Explain how this diagram shows that energy transfers are inefficient. *(3 marks)*

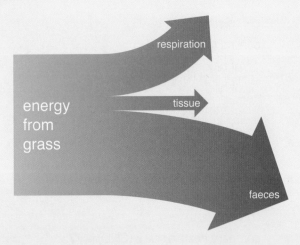

respiration

energy from grass

tissue

faeces

34 Explain why farmers might keep their cattle in heated barns during the winter. *(2 marks)*

35 a) State three ways of making the production of meat for food more energy efficient. *(3 marks)*

b) State one positive and one negative aspect of making meat production more efficient. *(2 marks)*

36 Explain why the number of people that an area of land can support depends on what food is produced on that land. *(2 marks)*

Answers online **Test yourself online** Online

Index